Simulating Nonlinear Circuits with Python Power Electronics

Shivkumar V. Iyer

Simulating Nonlinear Circuits with Python Power Electronics

An Open-Source Simulator, Based on Python™

python™

Springer

Shivkumar V. Iyer
Ontario
Canada

This book is based on the free and open source software Python Power Electronics hosted at the website http://www.pythonpowerelectronics.com
The software has been released under the BSD 3.0 license which be found in the License file with the software or at http://opensource.org/licenses/BSD-3-Clause

ISBN 978-3-319-73983-0 ISBN 978-3-319-73984-7 (eBook)
https://doi.org/10.1007/978-3-319-73984-7

Library of Congress Control Number: 2017963531

Printed on acid-free paper

This Springer imprint is published by the registered company Springer International Publishing AG part of Springer Nature
The registered company address is: Gewerbestrasse 11, 6330 Cham, Switzerland

To Rakhat

Acknowledgements

I would like to thank Prof. Bin Wu and Prof. Bala Venkatesh of the Department of Electrical and Computer Engineering, Ryerson University, for their interest and support in the development of Python Power Electronics. I would also like to thank Dr. Dan McGillivray who was the Executive Director of the Centre for Urban Energy, Ryerson University, for the many conversations we had related to the potential of the software. I would like to thank Mukundan R. of the National Institute of Industrial Engineering for his constant advice and suggestions. I would like to thank Zoubin Zarin for his advice in developing the software and with publishing this book. I would like to thank Janamejaya C for being a loyal follower of the project and his constant encouragement. I would like to thank the Python community and its many active members for all their support and for having me as a speaker in PyCon Canada 2016 and PyCon Canada 2013.

Shivkumar V. Iyer

Contents

Chapter 1
Introduction

Abstract This chapter introduces the concept of simulation by describing its usefulness with a few general engineering examples. The chapter then describes the current state of the power system and the recent changes that have occurred along with the changes that are expected in the future. The chapter describes the challenge of tackling climate change with renewable energy and the recent advances in wind and solar energy. With this background, the chapter introduces Python Power Electronics and describes the usefulness of a free and open-source circuit simulator and building a community of power engineers.

Keywords Simulation · Modern power systems · Renewable energy · Climate change · Smart grids · Power quality · Open-source technology

1.1 Concept Behind Simulation

Simulation from its basic definition is the imitation of an actual process. In modern times, the software definition of simulation is also available—the representation of the behavior or characteristics of one system through the use of another system, especially a computer program designed for the purpose. Though most engineers spend a significant amount of time simulating the systems they study, simulation as the above-defined concept is something almost everyone has been exposed to at some point of time. Simulation by its definition is when a real-world process is repeated. Let us look at some examples of simulation that we never stopped to think about.

For most of us who were born when computers were ubiquitous household machines, the first form of simulations is computer games. A computer game may be a car racing game, a war or fighting game, playing chess against the computer, or one of the strategy games that have become very popular recently. Almost all of these are imitations of real-world events. Behind the fancy graphics and the celebrity status characters lies a very complex algorithm that creates the environment where

S. V. Iyer, *Simulating Nonlinear Circuits with Python Power Electronics*,
https://doi.org/10.1007/978-3-319-73984-7_1

gamers find themselves constantly challenged. The environment changes to make the game tougher for the gamer in various forms - stronger enemies, tougher road conditions to drive on or more complex chess moves. These are examples of how a computer program has been designed to provide an entertaining platform for a user and also the program adjusts to user inputs not only in allowing the user to navigate the software but in making the platform more challenging for the user.

As the time comes for younger adults to get their driving licenses, they are quite often exposed to driving simulators. Unlike the video games, these have a more practical purpose in providing a new driver with necessary skills and also offering advice on road safety and illegal manoeuvres. Nowadays, there are some countries in the world where a driving license can be issued on passing a virtual driving test on a simulator rather than a road test. This shows how greater faith is being placed on simulators where the judgment of an experienced professional was the only one that was trusted. A more advanced form of simulator is the flight simulator made available to pilots during training. However, these are rarely accepted as the only form of training due to the critical nature of the skill to be acquired.

As adults, most of us have used simulators without knowing it, for example tax planning software. Almost everyone uses it when the time comes to file our tax returns. How much would we be paying in taxes? How much would we save by investing in Plan A and how does that compare with Plan B? Which expenses are tax deductible? Does your place of residence bring you tax benefits? Does your nature of employment entitle you to deductible expenses? If anyone reading this book has never used a tax planning software, you probably are paying way too much in taxes. In this case again, there is a fairly complex program that asks users for every detail that could affect their taxes and calculates the tax owed.

One of the most complex forms of simulation that have not yet been fully mastered is that of weather forecasting. With the mobile phone in everyone's hand, almost everyone looks at the weather forecast before stepping out of the house. Weather forecast is incredibly challenging because it is very strongly dependent on location - coastal versus inland, tropical versus temperate, equator versus sub-Arctic. And weather forecast still is an incredible challenge. The most powerful form of weather prediction is the storm watch. This is where a storm is tracked from the time it builds up to the time is finally dies. The number of variables that are involved in weather forecasting is simply mind-boggling.

With the above background on examples of simulators that everyone is exposed to, let us examine the specialized simulators that this book is all about. A simulation is a powerful and convenient way to monitor and understand a physical process. Simulation involves representing the physical process in the form of mathematical equations which is called a mathematical model and solving them over time. The inputs to the mathematical model are the factors which affect the physical process, and the outputs of the mathematical model are the physical quantities that are of interest for a variety of reasons—efficiency, safety, endurance, and many others.

1.2 The Reason for Simulation

There a number of reasons why engineers will want to simulate the system that they are designing, and we shall examine them one after the other. The simplest reason for simulating a physical system is because computers powerful enough to perform simulations are now ubiquitous. Fifty years back, in order to run a computer program, an engineer would have to buy processing "time" on a mainframe. This processing time was quite expensive, and therefore physical systems were simulated when building a prototype was very expensive. Nowadays, a reasonably powerful computer can be purchased off the shelf, and engineering software can be installed on almost every machine. So the answer to the first question "Why do you want to simulate a physical system?" is "Why not?".

To describe the other advantages, let us progress gradually from simple cases to more complex cases. If an engineer was designing a system that was fairly simple say for example a pendulum or a battery operated miniature motor, it is possible to skip the simulation step and by trial and error arrive at the final design. However, every design needs parts and supplies to be procured as well as structures fabricated. A change in the design will only result in repeated procurement of new supplies and parts and modifications to the structures or completely new structures becoming necessary. A simulation of these simple systems could significantly reduce the number of designs if not produce a working design at the first step. The result is lower time to finalize the design, lower cost of the project, and less waste.

Now for a more complicated case, let us suppose that an engineer needs to design a power supply. This design requires a detailed understanding of the power requirements of the intended load, the topology of the power supply that will achieve such quality of power, and finally a method to control devices such that this power quality is always achieved. In such an engineering project, there are several components and variables that need to chosen appropriately and furthermore, the choice of one affects the other. For example, a filter at the output of the power supply can be chosen such that it ensures that the power provided at the output is of a desired nature. However, this filter might make the control of the power device extremely complex or even impossible. If this engineering project were to be directly realized at the prototype stage, it would be very time-consuming, involving several changes and possibly even failures that could result in damaged components. On the other hand, a simulation would help to eliminate the designs that are completely infeasible. The final implementation stage could then require minor modifications as the simulation may not have taken some physical phenomenon into account. Simulation has thus helped to significantly simplify the design process.

Now to progress to an even more complex case and this time a non-mechanical one. Let us suppose that an engineer was to design a cooling system for an equipment. This cooling system could be a combination of heat fins on the outer surface and a network of pipes carrying cooling fluid. In such a design problem, trial and error method is almost ruled out as there are numerous variables to be considered. To begin with, a fairly detailed heat dissipation map of the equipment to be cooled has to be developed. The thermal simulation in this case will need to consider the worst case of heat emission in every part of the equipment along with the existing heat dissipation to the outside environment to determine the temperature pattern throughout the equipment. The objective is to determine the presence of heat spots which could cause damage to the equipment. After the thermal load of the equipment has been accurately determined, the cooling system has to be designed and repeatedly examined for change in the temperature map of the equipment. It would require several iterations before the cooling system is found to be adequate, and the design is finalized.

Now let us progress to a massive engineering project. Let us consider the design of a bridge. This would be an example of a huge civil engineering project. In such a case, simulations would need to be performed for a number of different problems—structural, thermal, corrosion, seismic (earthquake resistant), and even financial. The criticality of the design requires extremely detailed models that are verified over and over again by independent sources with different software. For example, from the structural point of view, vibration analysis could be performed to judge the stress on the bridge. The possibility of the bridge withstanding an earthquake will add a completely new dimension to the elasticity required from the bridge. Giant engineering projects such as these have a huge development cycle, and every possible aspect of the design needs to be analyzed and tested.

As can be seen from the build-up of cases, the need for simulation progresses from being desirable to absolutely essential. The advantage of simulation is the ability to repeat analysis for a number of conditions ranging from the best case to the worst case without causing any physical damage or destruction. The main challenge in ensuring that a simulation is accurate and reflects the physical system is to be able to model the physical phenomenon that is to be studied. Once a simulation model is validated and tested, it can be used repeatedly for examining physical systems of a certain category. The objective is to improve on the initial design that has been prepared using mathematical formula. Such a design could have fundamental flaws, could violate certain limits particularly in complex systems or may sustain damage in extreme conditions.

The process of simulations have also undergone significant changes. In the beginning, performing a simulation implied writing a program that would solve all the mathematical equations that describe the physical phenomenon. Such a process was in itself fairly tedious and error prone which in turn implied that unless the design was formidable to implement physically, a simulation may not be worth the effort due to the complexity. However, nowadays, there is a simulation software for every purpose—mechanical, chemical, electrical, financial. This specialized software has elaborate user interfaces that require data about the system being simulated in a particular format and generates the necessary equations and solves them. Output data is available in a number of formats—spreadsheets, plots, and many more. The process of simulation has therefore become simplified to the extent that a user need not bother about the mathematical model that needs to be formulated and solved.

Most universities in the world expose engineering students to simulation software at a fairly early stage. Besides programming is quite often mandatory at different levels which makes the use of simulation software easier in case these software required some basic programming capability. At the graduate level, almost every thesis and assignment requires detailed simulation which makes the modern engineering workforce fairly skilled at the process of simulation. Simulation is a core component of the engineering process and will continue to grow as time progresses. The demands from simulation in turn are increasing as the need to model physical processes to a greater detail is increasing.

1.3 Simulation in Power Electronics and the Challenges

With the above background on simulation in general, let us now dive into the topic of this book—simulation for power electronics. A number of simulators exist for power electronics, most of them being commercially licensed. Most text books on power electronics now have simulation examples provided along with them which students can use to help learn theory. Most of the popularly used converters have simulation models that can either be freely downloaded as packages with software or can be obtained as a supplement of textbooks. Simulation has become a very effective technique in testing control strategies on converters. Simulations also help with loss calculations and determining the efficiency of converters.

The energy industry has been undergoing a major transformation in the past couple of decades, and far more changes are expected. The biggest factor in the power system is now renewable energy, primarily wind and solar, but a whole host of others as well - tidal energy, geothermal energy, biomass, and several others. There are a few countries in the world such as Denmark that have already achieved over 50% of renewable energy penetration. Increasing renewable energy in the power system is great news for the environment amid concerns of global warming and climate change. However, for the power system operator, increasing renewable energy is a cause of concern with respect to power system reliability. This is for the simple reason that power is now being generated in parts of the system that were never expected to be generation stations.

The conventional power system consisted of remote power plants—thermal (coal or nuclear) or hydroelectric—that would supply electricity to load centers—urban, rural and industrial. The supply of electricity would require an entire network of transmission lines, substations, and distribution lines with the voltage being transformed to different levels to ensure efficient transmission. The power system operator had to ensure that power demand from the load centers was met by adequate generation and that transmission and distribution lines were never overloaded. Another aspect was to control the voltage level at different parts of the power system to ensure safety and continuity of power. With renewable energy generation, power is now being generated all over the power system—the load centers, the distribution system, and the transmission system. An example of this is the rooftop solar photovoltaic panels that are generating power at the customer load center or the wind farm that is connected to the transmission system at a substation.

The biggest challenge with integrating renewable energy is the intermittent nature of the sources such as solar photovoltaics and wind. A significant proportion of renewables can therefore disrupt the operation of the power system. Cases have been reported about wind farms in rural locations causing voltage fluctuations since changing wind speeds and the associated fluctuating power alter the voltage at the system where the farm is connected. A number of problems have been reported with renewables, some of which have fairly simple solutions while some have complex and expensive solutions. The pressure to decrease fossil fuel-based electricity generation with renewable energy-based generation therefore needs viable solutions to interconnection issues experienced with renewable energy.

On the other hand, the loads and appliances have also significantly changed over the years. Increased automation and energy conservation in appliances have made them more efficient but have made them more sensitive to voltage fluctuations. Industrial processes have now been automated to remove the need for human operators. This process of automation has reduced their immunity to voltage changes. Many industries are located remotely for a number of reasons, and this implies that the voltage supplied to them is not of the same quality as an urban customer. Voltage fluctuations that would normally have been tolerated by the older equipment will cause new equipment to either reset or fault. Most utilities typically guarantee the voltage supplied to a customer between $\pm$3–5%. Nowadays, voltage fluctuations within this range are sufficient to cause equipment to malfunction.

In most of the above cases, power electronics plays a major role. Most renewables such as solar and wind have power electronics interfaces to extract the maximum possible energy from them and convert the energy to conform to grid regulations. Automated processes quite often have power electronics interfaces, an example of which are variable speed drives. Most voltage conditioning and power factor correction equipment also have power electronics converters. Therefore, to fully understand the operation of these modern power systems, it is necessary to be able to model these power converters along with their controls. A simulation study of a modern power system will now have to include the simulation of power converters to be able to examine the impact of renewable sources.

An example of such a simulation study is as follows. Suppose a wind farm connected at a rural location at a transmission voltage level of 120 kV is the focus of attention as industrial customers in the vicinity have reported fluctuating voltage. A number of scenarios are possible. The wind farm has been inappropriately sized due to which power ramps can cause the voltage at the point of connection to fluctuate out of bounds. Another may be that the wind farm is not excessively sized, but functioning of the wind turbine controls are such that the transients during power ramps force the voltage out of bounds. It is also possible that the voltage may not be fluctuating out of utility specified bounds, but the industries have processes that are extremely sensitive to any voltage changes. In scenarios like these where a lot of finger pointing occurs, simple measurements at different buses to determine how voltage changes may not shed sufficient light on the problem. A detailed simulation study to examine every transient in that part of the power system may need to be performed.

The simulation of power electronic converters to verify control strategies and estimate their efficiency is now fairy well established. A number of simulation software can be used for this purpose. Software that can simulate distributed systems such as a segment of a power system with a solar farm or a wind farm are on the other hand a specialized field. Many simulation software perform the task of processing distributed systems by either performing a traditional power system type load flow analysis or by approximating them and performing an energy balance. The disadvantage of this method is that the transients associated with the control loops of the power converters are not captured. These simulation techniques are still effective in determining power flows, efficiency, and potential overloading, and therefore, their use in the planning stage is strongly entrenched. However, the power quality issues related to cycle-by-cycle fluctuation needs a more detailed approach.

The approach that would be effective in examining power quality issues is to perform a detailed transient analysis. This transient analysis will build a mathematical model of the power system from the fundamental network laws, namely Kirchhoff's Voltage Law and Kirchhoff's Current Law. The mathematical model will be solved taking as inputs the instantaneous values of available power system voltages, switching patterns of power converters, and generating as outputs the currents in all parts of the power system. This form of transient simulation is typically employed for smaller circuits as the computational burden for a small system is not significant. However, for a power system, performing transient analysis is a challenge.

The first challenge is in developing accurate mathematical models of the power system in the presence of power electronic converters. The second challenge in solving the equations pertaining to the model and handling the amount of data generated by the solution. The second challenge is of critical importance when a simulation needs to be executed for a long time duration as the data generated by simulators quite often causes even advanced computers to exceed their memory limits. A simulation of a complicated system with multiple power converters could require several days to complete, and in the case of many software, the intensive nature of the computations results in all the resources of the computer being consumed by the simulation software. This leads to inefficient use of computational resources, memory, and eventually in longer development times for projects.

To address these challenges, Python Power Electronics was developed. The next section will describe how the software developed and the current state of the project.

1.4 Python Power Electronics

The origins of this software go back to when I was a graduate student and was simulating power electronic converters for my Master's and later my Ph.D. thesis. I was working on the topic of microgrids and that meant multiple converters in a single system. To speed up the simulations, I would solve the equations for the circuits in C++. The advantages were significant. Simulations were now several times faster, and I could run multiple simulations simultaneously. Moreover, by using a Unix-based operating system, I could further improve memory management and avoid my computer running out of resources. In addition to simulations, I was writing manuscripts, preparing presentations, and finally writing my thesis. The only disadvantage was that all the equations for every circuit had to be completely programmed in the simulations which increased the initial development time.

After graduating, I thought of means of improving the above simulator to make it more automatic. If the user had to write all the equations for every circuit, very few would ever use the simulator. Therefore, like commercial simulators, the user need only provides a schematic and the simulator should be able to generate the necessary equations and solve it. The project began as a blog that can be found at:
http://www.pythonpowerelectronics.blogspot.com
The user interface for designing circuits was conceptualized. Circuits could be designed in spreadsheets that were saved as Comma Separated Value (.csv) files. The simulator was designed to read the connectivity information of the circuit and generate the mathematical model. After programming an equation solver, the simulator was able to solve passive linear circuits.

After building a basic simulator that could simulate passive linear circuits, the software was released free and open source on SourceForge at:
http://sourceforge.net/projects/pythonpowerelec/
After another year of development and releasing several versions of the software, the simulator was able to solve nonlinear circuits with power converters. The development of the simulator has been extensively documented in the blog. Eventually, a dedicated Web site was chosen for the simulator and can be found at:
http://www.pythonpowerelectronics.com
This Web site contains documentation on using the software, download links for the software, and case studies on how sample circuits have been simulated.

The first question—why Python? This will be described in the next chapter in detail, but at this point, it should be mentioned that Python is a free and open-source high-level language that is being extensively used especially in the scientific community. The second question—why open source? I have been a user of open-source software for over a decade—Linux for the operating system, Python/C for programming, Latex/LibreOffice for documentation. I am a strong advocate of open-source software as I believe it leads to better code in the long run and a greater user base. One of the greatest problems faced as a power engineer has been the lack of good open-source simulators since almost all the simulators commonly used by students are commercially licensed. Over the years, as I have moved between universities and companies, I have used numerous simulation software which has involved repeating simulations in different platforms and duplicating work over and over again. My hope is to be able to build a simulation software that can be used by power engineers like me to develop simulations models over the long run without bothering about software licensing.

As stated in the previous section, one of the biggest challenges faced by simulators when solving distributed circuits with multiple power converters is the burden of handling large amounts of data. Typically, most software maintains simulation data in volatile memory (RAM) as this ensures faster processing. However, once the simulation progresses for a long duration, the only option is to temporarily write data to the hard disk and free up RAM space for future processing. In Python Power Electronics, a minimal amount of data is maintained in the RAM and output data is in general written to the hard disk. The data maintained in RAM through the in-built cache memory of Python are for the current and the previous instant of simulation. Though this may slow the processing for smaller systems, for larger systems, this prevents any memory overflows and makes the simulator stable.

Python Power Electronics has evolved considerably over the past four years since its first launch. It has been tested for a number of circuits with different topologies of power converters. The software has been modified every time a bug has been found, and a new version has been released. The complexity of the circuits and the number of power converters have been continuously increasing with the objective of simulating a power system with a wind or solar farm. However, besides also being a software that has an industrial or research application, another significant aspect in the development process has been the learning experience. Developing this circuit simulator has provided a great wealth of information about nonlinear circuits and using network laws to solve them. To compile all the lessons learned over the past four years and which are currently in blog posts, into a book that could be of interest to students and practicing engineers in power electronics has been the motivation behind writing this book.

The next section will provide an outline of the book and the contents in every chapter.

1.5 Outline of the Book

The book consists of seven chapters and can be viewed in two parts though such a division is not actually made. The first four chapters describe the outer layer of the software or the user interface and in particular examine how the simulator processes the user interface. The last three chapters on the other hand examine the core simulation engine and how circuit analysis is performed. The objective of the first part is to describe to a user how the simulator is to be used by providing sample circuits, programs, and finally a complete case study. The objective of the second part is to describe to the user how the simulator solves the circuit, and since this entire operation is invisible to the user, the process is described as a concept rather than through code.

Chapter 2 provides an overview of the Python programming language. As stated in this chapter, the purpose is to provide the user a short introduction to Python. For a detailed tutorial, a reader is recommended to either read a book on Python programming or follow an online tutorial on the internet. The purpose of the chapter is to enable the reader to understand the code segments that will be provided in the subsequent chapters that deal with user-defined control functions and the case study.

Chapter 3 describes the interface that the simulator uses to interact with the user. The chapter describes the philosophy behind choosing spreadsheets as the mode of extracting information from the user. Spreadsheets are used by the user to enter simulation parameters, circuit schematics, parameters of the components in the circuit schematics, and also the structure of control functions. The chapter describes how the structure of every component class in the simulator library and how the data entered by the user is processed by each component class. The chapter also describes the concept of how classes are instantiated for every component found resulting in objects and how these objects are referenced by the simulator. The chapter describes the execution flow in the simulator and how the simulator processes the data provided by the user and makes it available to the core simulation engine. The chapter does not describe how user-defined control functions are processed as the whole of Chap. 4 is dedicated for this purpose.

Chapter 4 describes how a user can write control functions for a simulation. Chapter 3 has described which of the circuit components can be controlled externally. Besides these controllable components, a control function need not perform a control action, but can instead be used to process simulation data or perform calculations. The chapter describes how the control functions have to be written as Python 2 files and specified in the simulation parameter spreadsheet. Every control function will have an interface to the simulation in terms of inputs and outputs, and this interface is described by a spreadsheet called a descriptor. Besides inputs and outputs, every control function can use certain types of variables that perform special functions. The chapter describes the importance of each type of control variable and how they are implemented in the simulator. The chapter describes how control functions are scheduled by the simulator using time events, and with an example, it is described how the simulator ensures that the control functions execute at the desired time instant. A simple example has been provided to describe how control functions can be interfaced with the simulation and also with each other.

Chapter 5 describes how a user can simulate a circuit with a power electronic converter. The example chosen has been a shunt connected three-phase VAR compensator realized using a two-level voltage source converter in a three-phase system. The voltage source converter consists of controllable ideal Switches that are turned on and turned off by pulse width modulation. The chapter describes how the user can build this simulation in stages such that every new subsystem added to the circuit can be verified. The chapter also describes how the user can write control functions with detailed examples of each control function in the simulation and also design the control interfaces through descriptors. Every stage of the chapter contains simulation results to show how the project develops. Through this example, every feature of the simulator has been described with details so that users can develop their own simulations.

Chapter 6 describes how the simulator processes the circuit schematics that the user enters in spreadsheets. The connectivity information is extracted from the circuit schematics in the form of nodes, branches, and loops. Nodes, branches, and loops are used to perform circuit analysis through loop analysis and nodal analysis which are described in the next chapters. The chapter describes through sample circuits, the algorithms used to determine the nodes, branches, and loops. The chapter introduces the concept of the LoopMap which is used for performing loop analysis in Chap. 7

and the concept of KCLBranchMap which is used for performing nodal analysis in Chap. 8.

Chapter 7 describes how loop analysis is performed in the simulator. The chapter describes how the matrix equation for performing loop analysis is generated from the LoopMap described in Chap. 6. A brief description is provided about how the matrices in this equation are transformed using row operations such that they can be solved by using numerical integration techniques. The chapter describes how loop currents and branch currents in the circuit can be mapped which allows for calculation of branch currents from loop currents and vice versa. The chapter describes with an example how time constants of branches of the circuit can make the simulation unstable and introduces the concept of a stiff loop. By providing a sample circuit and its corresponding LoopMap, the chapter describes the need to isolate stiff loops so as to be able to simulate a circuit. With this example, the concept of loop manipulations is described, and with advanced examples, the effectiveness of the procedure is described. The chapter describes the limitation of loop analysis with another set of examples and therefore the need for nodal analysis.

Chapter 8 describes how nodal analysis can be used to determine the currents through stiff branches (that have a very low time constant) in the circuit. With the example of a simple buck converter, the chapter describes how loop analysis is insufficient in determining the conduction of power devices during switching events. The chapter then describes how nodal analysis can be used effectively in determining how power devices conduct and the transfer of current from one device to another. The chapter introduces the concept of events and how the matrix equations for the circuit will be constant until an event occurs. The chapter finally describes the logical flow of processes in the simulator as it performs loop analysis and nodal analysis one after the other.

Chapter 9 will conclude the book by highlighting the advantages of the simulator and the future development intended in this project.

1.6 About the Book

One of the prime motivators for writing this book has been the lack of documentation available with Python Power Electronics as a software. The documentation available is on the Web site and the blog, but however, a detailed user manual does not exist. Instead of preparing a user manual, I decided to take a step further and write about the simulator in the form of a detailed book describing not just examples for users to be able to simulate their own circuits but also to describe the theory behind the simulator. The simulator is completely free and open source, and this would enable interested students to combine the theory in this book with the code in the simulator to better understand circuits. As stated before, one of the greatest advantages of this project has been the learning experience with respect to understanding how power electronic circuits function. As a graduate student of power electronics, I found very few resources to understand advanced power electronic circuits through simulations. With the energy industry undergoing significant transformations, the expectation from new engineers entering this sector has been continuously increasing. It is my hope that through open-source software and community-owned projects which promote sharing of knowledge, the energy industry gets well-prepared and knowledgeable engineers.

Chapter 2
Introduction to Python

Abstract This chapter introduces the Python programming language briefly so as to serve as a quick reference for the later chapters. This chapter describes the features of the programming language and also the different types of objects and their associated functions. This chapter does not serve as a detailed tutorial on Python programming but is a quick reference for the material in the book.

Keywords Python programming language · Object-oriented programming · Integrated development environments (IDEs) · Compilers · Operating systems · Open-source software

2.1 Introduction

The simulator has been developed using Python and in-built modules. The choice of Python language as opposed to other high-level languages such as C, C++, or Java was to take advantage of the fact that Python is being adopted by the scientific community for a large number of applications. Furthermore, many of these applications are being developed in research laboratories and universities and a significant number of them are completely open source. An open-source application has the greatest advantage that it can be developed and modified by several independent groups simultaneously resulting in better software and greater community engagement. Reporting of bugs and the subsequent bug fixes results in cleaner code. As of now, the circuit simulator is not completely independent as an application. To use the simulator, a user will need a few additional software which will be described in this chapter.

The circuit simulator Python Power Electronics is completely free and open source. To trace back to the very origins of the simulator, the reader is encouraged to read the blog:

http://pythonpowerelectronics.blogspot.ca

S. V. Iyer, *Simulating Nonlinear Circuits with Python Power Electronics*,
https://doi.org/10.1007/978-3-319-73984-7_2

The simulator began as a hobby and with mere blog posts. After a few months, with a basic working version of a simulator, a project was started in SourceForge:

http://sourceforge.net/projects/pythonpowerelec/

In November 2015, the circuit simulator and all associated documentation have been made available at the Web site:

http://www.pythonpowerelectronics.com

This Web site contains links to download the software, documentation on using the software, and a blog which is a collection of case studies where the simulator has been used to study different circuits.

It is highly recommended that a user that wishes to use this simulator extensively should try to hack into the source code as this would help in debugging simulations and a better understanding of how circuits are solved. This chapter will provide a basic introduction to the Python language. For a detailed and in-depth tutorial, a user is recommended to read a book [1] on Python or go through the documentation in the Python Web site (http://www.python.org). Several free tutorials of various degrees of user expertise are available on the Internet on Python and its modules. The purpose of this chapter is to enable the user to understand the implementation of circuit analysis functions in Python in the following chapters and also to be able to write control functions. This chapter will provide a basic overview of the Python objects and functions used in the circuit simulator and also used later in Chaps. 4 and 5 while describing how control functions can be implemented in simulations.

2.2 Overview of Python

Python was created in the 1990s by Guido van Rossum at Stichting Mathematisch Centrum, and he now holds the title given to him by the Python community "benevolent dictator for life." The focus behind this language was simplicity, improved code readability, and dynamic programming capabilities. It has been adopted by many groups all over the software world—scientific computing, Web development, database management, and many more. What has increased the attractiveness of Python is its being free and open source which enables anyone to build an application using it for any purpose. This in turn has driven the development of Python Power Electronics. This section will describe in brief how Python is different from other high-level languages and why Python was chosen as the software for development of this circuit simulator.

The reader is encouraged to look up the Zen of Python which dictates the guiding principles behind the design of the programming language. One quick read will make it obvious that one of the core philosophies of Python was a language where the code is easily readable. As programmers, we are encouraged to indent blocks of code to improve readability, e.g., a block of code within a loop, within an if-then-else condition, within a function or any other purpose where a block of code has a special significance. In Python, indentation is mandatory. There are no open and close braces ({}) like there is in C/C++ nor are there Begin/End statements. The indentation is the way the compiler knows which block of code comes within a block like a loop or a function. Similarly, when the indentation returns back to normal, the compiler knows the block has ended. The indentation automatically improves readability of code as now code blocks are easy to view and code is easier to debug.

Python is a dynamic language. This has a very long and elaborate description for which the user should seek documentation on the Web site or [1], but what will be provided here is the short version of it. In Python, there is no need to declare a variable as in C/C++ and in many other languages. A variable is created when it assigned a value. Additionally, everything in Python is an object—every variable, every class, every function, almost everything else. An object is a collection of data and functions that operate on those data. In Python, if you assign a string to a variable, it is not just data equal to the string that is created, but an object with the data being the string and a host of functions that can operate on this data are automatically available. More on this later. The advantage of this is that the user does not need to create functions for basic operations on the data as these are automatically provided by the compiler when the variable of that data type is created. As an extension of data types, another useful feature in Python is the ability to create complex embedded data types. This feature is used heavily in this circuit simulator. As an example, it is possible to create a variable that is a list whose elements are in turn a combination of lists, floating point numbers, and integers. Since an electrical circuit is dynamic system and even more so a power electronic circuit which is nonlinear, the ability to define variables whose very structure and content can be changed is a great asset.

Another feature of Python is the limited syntax which is a part of the philosophy of the language. For example, in Python, a loop can be achieved using two possible constructs—a for loop or a while loop. In contrast, C/C++ have for loops, while loops, do-while loops, do-until loops, and maybe even more. The philosophy of Python is if a set of tasks need two or three constructs at most, let there be two or three constructs at most. The advantage of such a method is that code is quite often similar even though written by programmers with very different backgrounds. This makes it much easier as a developer to hack into and build on code already released open source. The simplicity of the language is also one of the reasons for it being used in diverse applications. Some of the most popular frameworks built on Python are Django (a Web development framework), NumPy, and SciPy (scientific and mathematical frameworks) and Matplotlib (a plotting software). For a full list of the software based on Python, visit the Python Web site or perform a simple Internet search. The applications of Python are simply mind-boggling, and in particular, the interest that the scientific community is beginning to show in adopting Python was one of the main reasons for choosing Python to build this circuit simulator.

2.3 Getting Started

The circuit simulator is built on Python 2 though Python 3 compatibility will be incorporated in the near future. At the time of writing the book, Python 2 is a requirement and therefore the user is recommended to install Python 2.7.x and ensure that all Python files related to the simulator are being compiled as Python 2 files. This section will describe the requirements for using the circuit simulator.

The first step is installing Python 2.7.x. A user has several options depending on the operating system being used. There are several integrated development environments (IDEs) available with Python. These are software where the user can create, write, and edit Python code and also compile them to view the result in a separate shell. Many IDEs are free, while some are proprietary—the complete list can be found at the Web site:

http://wiki.python.org/moin/IntegratedDevelopmentEnvironments.

The user is free to experiment with a few of them until one of them is found suitable. Reviews for these IDEs can be found in many forums. The preferred way of using the simulator is to use it in a Unix-based operating system with Python invoked from the command line in a shell such as BASH (Bourne Again Shell). The Python programs can be edited using any editor Vim, gedit, etc. Such a combination makes the simulator lightweight and eliminates problems like software crashing or hanging. However, it is useful to have an IDE with a Python shell as this shell allows you to execute Python commands and view the results immediately which can be useful for debugging.

The simulator does not have a graphical user interface (GUI). All inputs to the simulator are either through the command line or through spreadsheets. Emphasis is placed on inputs through spreadsheets as in that case, inputs will not have to be repeatedly entered by the user as would be the case in command line. For this reason, it is necessary to have a software that can edit a spreadsheet. Such software exists in every operating system, and there are also free and open-source options. The preferred software is LibreOffice https://www.libreoffice.org/. LibreOffice is free and open source and has a spreadsheet software called LibreOffice Calc. No comments will be made on the overall performance of this or any other software. For the purpose of the circuit simulator, LibreOffice can create and edit spreadsheets and save them as Comma Separated Value (.csv) files.

Finally, the output of the circuit simulator is data written to a text file. The output data are the measured outputs of different meters or other variables designated by the user to be written to the output data file. Each data item is available as a column in the data file separated from other items by a single white space. The first column is the time instant of the simulation. Each row of the data file corresponds to all the measured data and other variables at a particular instant of time in the simulation. The plotting software needs to be able to plot columns with respect to each other. The preferred plotting software is Gnuplot which is also free and open source. There are many other plotting software, and some may need the data to be made available in another format. The biggest advantage of using Gnuplot is that the waveforms can be observed as the simulation proceeds by refreshing the plots. In this manner, the simulation and plotting output data can be independent tasks. It should be noted that plotting typically becomes a major computational burden for simulation of large circuits for long time durations when the output data files can exceed sizes of hundreds of megabytes. Gnuplot has been found to be a software that can handle such large amounts of data in the best possible manner.

With a brief background on the software the user should become familiar with, the next section will start with an introduction to the Python language.

2.4 Integers, Floats, and Strings

These are the basic data types used in the circuit simulator and also in control functions written by the user. Despite being basic data types, in Python, variables of these data types are objects. This implies that besides the value of the variable, the variable contains other associated data and functions that act on the value of the variable.

For creating a variable of data type integer, all that needs to be done is to assign an integer value to a variable. The following statement:

```
a = 4
```

will create an integer type object with value 4. To understand the concept of an integer object, do the following in your IDE after the above statement:

```
Type a. and then press <TAB> key.
```

Most IDEs will list all the functions and data that are associated with this object since "." is the delimiter that allows us to access the member associated with an object. The integer object a will have the following member data and functions:

```
a.numerator
a.denominator
a.real
a.imag
a.bit_length()
a.conjugate()
```

To know what these items are, enter:

```
print a.conjugate.__doc__
```

This will print out the documentation related to the member function conjugate of the integer data type a which is:

```
Returns self, the complex conjugate of any int.
```

This gives an example on how Python treats any variable as an object. Even something as simple as an integer comes with member functions and member data which though in the context of the circuit simulator is unnecessary. In the simulator, integers are the indices of vectors and matrices. Also, in a loop, integers are used to increment the counter of the loop.

Similar to integers, a floating point number is created by assigning a floating point number to a variable. As stated before, when a value is assigned to a variable, an object of that data type is created and assigned to the variable. Therefore, if the variable a which was assigned an integer value is assigned a floating point value as:

```
a = 4.3
```

The integer type object assigned to a is destroyed, and a floating point object is assigned to a with the value of 4.3. As with the integer type, this object has several member functions and member data. In the simulator however, a floating point number typically is used for most physical quantities. Integer and floating point variables can be a part of mathematical operations—addition, subtraction, multiplication, and division. The Python documentation is a good starting point for all the possible mathematical operations that can be performed on integers and floating point numbers. The circuit simulator performs a whole host of calculations of integers and floating point numbers that are essential for solving equations.

A variable of type string is created by assigning a string to the variable. For example,

```
a =''hello there''
```

A string object however has several useful member functions. In the simulator, any input from the user through either the command line or a spreadsheet is through strings. Therefore, the handling of strings is fairly important in any program that has a user interface. Two useful member functions in any string object are lower() and upper().

```
a.upper()
```

will produce "HELLO THERE" as the function converts the string to uppercase. The function lower() in contrast converts the string completely to lowercase. Another extremely useful function is the split() function. This function will split the string using the delimiter provided (if no delimiter is specified, it is assumed as a single space) to produce a list of strings separated by the delimiter. For example,

```
a.split()
```

will produce ["hello", "there"] with the default delimiter being a single space. On the other hand,

```
a.split(''o'')
```

will produce ["hell", "there"] since the delimiter is now the character "o". The delimiter need not be a single character and can be a string. If the delimiter is not found in the string, the entire string is returned back. The split() function is used extensively in the simulator as it helps to search for patterns in a string. For example, if looking for the extension of a file, split(".") can be used to separate the file name and the extension. The reader is encouraged to check out the member functions of a sample string in an IDE and try out some of them. Additionally, it is possible to concatenate two strings by "adding" them together.

```
''hello'' + '' there''
```

will produce "hello there". This feature is also used quite often in the simulator when it becomes essential to output to the user a variable in a format such that it can be understood by the user.

It is possible to convert one form of variable to another. For example,

```
str(4.3)
```

will produce a string "4.3". Similarly, this can be converted back to a floating point number by

```
float(''4.3'')
```

However, the operation,

```
int(''4.3'')
```

will produce an error as the string when converted into a number is a floating point number. It is possible to convert a float to a string and vice versa,

```
int(float(''4.3''))
```

will produce 4. Similarly,

```
str(float(4))
```

will produce "4.0". This conversion becomes essential when trying to read numbers from either command line input or from a spreadsheet. The values read from command line or spreadsheets are strings, and they need to be converted to either floating point numbers or integers. There are other numeric data types in Python such as "long" and "complex". These are not used in the circuit simulator and therefore are not described here. However, these can be found in the Python documentation or in any programming book based on Python [1]. The later sections will provide examples on how the different data types can be used and manipulated particularly using programs that are relevant to the simulator. For a more general understanding of data types, a reader is recommended to follow a course in Python.

2.5 Lists and Tuples

Both lists and tuples are extremely powerful built-in types provided in Python. Lists are used extensively in the circuit simulator though tuples also have some useful properties. Both lists and tuples are collections of objects—integers, floats, strings, or even other objects. Let us consider a list x and a tuple y.

```
x = [1, 2, 3, 4]
y = (1, 2, 3, 4)
```

As can be seen, a list is defined with square brackets [] while a tuple with round brackets (). The examples above are fairly simple. Consider the following:

```
x = [1, ``Monday'', 4.7, ``September'', \
                [1, ``Active'', 2232.65]]
y = (1, ``Monday'', 4.7, ``September'', \
                [1, ``Active'', 2232.65])
```

As stated before, the objects in a list can be in any form—they do not have to be of the same type, and they can also be other lists or tuples. This feature makes lists and tuples extremely powerful as it let you collect all kinds of information together. The difference between lists and tuples is that tuples are immutable (cannot be changed) while lists are mutable (can be changed). To look into this further, try to create the list x and tuple y above in an IDE. Use the . operator with them to examine the member functions and data available with these objects.

There are some extremely useful functions with the list object that are frequently used in the circuit simulator. One of them is the append function which is used as:

```
x.append(``Interest'')
```

which will add "Interest" or for that matter any object specified within the function to the list x to produce:

```
x = [1, ``Monday'', 4.7, ``September'', \
                [1, ``Active'', 2232.65], \
                ``Interest'']
```

The extend function is also used frequently with lists.

```
x.extend([``Quarterly'', 512.23])
```

will use the parameter (usually another list) specified in the function to extend the list x and will produce:

```
x = [1, ``Monday'', 4.7, ``September'', \
                [1, ``Active'', 2232.65], \
                ``Interest'', ``Quarterly'', \
                512.23]
```

In contrast the append function will merely add the object specified in the function. So the difference between the functions would be that append with the same argument:

```
x.append([``Quarterly'', 512.23])
```

will produce:

```
x = [1, ``Monday'', 4.7, ``September'', \
                [1, ``Active'', 2232.65], \
                ``Interest'', \
                [``Quarterly'', 512.23]]
```

As can be seen, the list object ["Quarterly", 512.23] has been added to the list x rather than extending list x with the list argument which is what the extend function does.

A list element can be accessed as:

```
x[3]
```

will produce:

```
''September''
```

It should be noted that list indices start at 0. Therefore, x[0] is the first element. There is another very handy functionality provided with lists. The last element of a list can be accessed in two ways by accessing the element using the length of the list or by using the −1 index.

```
x[len(x)-1]
```

will give the last element of the list at any time which will be:

```
[''Quarterly'', 512.23]
```

It is to be noted that the above result takes into account the append statement. At the same time:

```
x[-1]
```

will produce the same result. Therefore, any negative index implies accessing the list elements backwards. Similarly, x[-2] will produce the second last element of the list x, x[-3] the third last element, and so on.

On the other hand, none of these functions are available in a tuple. It is not possible to add or extend a tuple. However, a tuple is faster to iterate through and therefore, if an object is known not to change, declaring it as a tuple allows for faster processing. Moreover, a tuple being immutable can be used as a dictionary key while a list cannot. Dictionaries will be dealt with in the next section.

2.6 Dictionaries

A dictionary is a data type in Python that maps objects to labels called as keys. Dictionaries are extensively used in the circuit simulator. A simple example of a dictionary is below:

```
a = {''resistor'': 100, ''capacitor'': [0.1, 0.1], \
                (100,0): ''position''}
```

A dictionary is a collection of ⟨key⟩:⟨value⟩ pairs within braces . For example, "resistor": 100 is a ⟨key⟩:⟨value⟩ pair. The key is a label for the item and has to be unique in the dictionary as the dictionary item is accessed with the item key. For example,

```
a[''resistor'']
```

would produce the value 100. The value associated with a key in a dictionary item can be any object—any number, string, list, tuple, or even a user-defined object. However, the key must be unique and immutable and can be either a string or a tuple as neither of these are mutable.

The advantage of using dictionaries is that if a collection of objects can be accessed by unique identifiers, making them dictionary items allows for fast access. Python has fast search functions that can retrieve objects that are dictionary items. The challenge lies in generating unique keys for all the dictionary items. The circuit simulator uses dictionaries for several purposes. A dictionary is used for storing the objects for every circuit component with their keys being the spreadsheet position of the components. A dictionary is also used for storing the branch information so as to be able to store loop information and associated information for performing circuit analysis. Using a dictionary removes the need to search for an object as would be the case if the objects were in a list. Moreover, a dictionary is not ordered. In the above example, the dictionary "a" has been defined in the order as shown above. But Python may find another way to order the dictionary items and may change the order as new items are added to it. The most important thing in Python dictionaries is the ⟨key⟩:⟨value⟩ mapping.

To create a dictionary, there are several methods which can be looked up in the documentation. The method used in the simulator is to use the key for the item. For example, let us suppose an empty dictionary is created:

```
a = {}
```

As stated before, the Python dictionary is not ordered. Therefore, the items do not have indices like in lists. To add the item with key "resistor",

```
a[''resistor''] = 100
```

will create the dictionary item. So then the dictionary will be:

```
a = {''resistor'':100}
```

To change the value of the dictionary item,

```
a[''resistor''] = 200
```

Accessing a dictionary item with a key that does not exist is an error. Therefore, the following statement:

```
print a[''capacitor'']
```

will produce an error as the dictionary "a" does not have an item with key "capacitor". To know which keys exist in a dictionary, a dictionary has the keys() function that returns a list of all the keys present in the dictionary. Therefore,

```
if ''capacitor'' in a.keys():
        print a[''capacitor'']
```

will prevent a key error. The function a.keys() will return the list ["resistor"] as "resistor" is the only key in the dictionary which has only a single item. To create the dictionary item with key "capacitor", the following two methods are possible.

```
a[''capacitor''] = [0.1, 0.1]
```

This statement will create an item with key "capacitor" and assign the list [0.1, 0.1] as a value to this item. However, in case an item with key "capacitor" exists, the above statement would delete the object assigned to the item and replace it with the list [0.1, 0.1]. If that is not what is intended,

```
if ''capacitor'' in a.keys():
        a[''capacitor''].append(1000.0)
else:
        a[''capacitor''] = [0.1, 0.1]
```

In this structure, if an item with key "capacitor" exists and if its value is known to be a list, the float 1000.0 is appended to the list while if the item does not exist, it is initialized to the list [0.1, 0.1]. In actual practice, dictionary manipulations may be fairly complex.

2.7 Examples of Python Code in the Simulator—Lists and Matrices

The previous section described the Python data types that are heavily used by the circuit simulator. The documentation on Python is vast, and it is recommended that the reader of this book spends some time in at least getting familiar with the scope of the programming language and the immense power that it provides the programmer. The fact that this entire circuit simulator has been written in basic Python shows how flexible the language is. However, the reader is encouraged to explore the additional modules available with Python such as NumPy and SciPy. This section will get the reader quickly acquainted with programming techniques in Python. This would be useful if the reader wishes to immediately write control functions as will be described in the later chapter.

Let us begin with matrices and matrix manipulations. To represent a vector, the simulator uses a simple list. For example, the input vector to the circuit which is usually the voltage sources in the circuit can be collected together in a list. In the case of a three-phase circuit with a three-phase voltage source, the input vector would be written as follows:

```
import math
va = math.sqrt(2)*120.0*math.sin(2*math.pi*60*t)
vb = math.sqrt(2)*120.0*math.sin(2*math.pi*60*t - \
        120*math.pi/180)
vc = math.sqrt(2)*120.0*math.sin(2*math.pi*60*t - \
        240*math.pi/180)
sys_u = [va, vb, vc]
```

In the above code example, "import math" imports the in-built math module available with Python. This module has useful functions such as $\sqrt{}$ (sqrt), sine (sin) or constants like π (pi). These objects (functions or constants) can be accessed as members of the math module. In order to generate the above source voltages for a particular time duration with a fixed time step, the code is as follows:

```
import math
for t_index in range(1000):
        t = t_index*10.0e-6
        va = math.sqrt(2)*120.0*math.sin(2*math.pi*60*t)
        vb = math.sqrt(2)*120.0*math.sin(2*math.pi*60*t - \
                120*math.pi/180)
        vc = math.sqrt(2)*120.0*math.sin(2*math.pi*60*t - \
                240*math.pi/180)
        sys_u = [va, vb, vc]
```

The for loop generates a range of values for the variable t. Since loops are a very critical part of the simulator and are in conjunction with lists, let us examine them together.

The function "range" used above is a very useful function in Python that generates a list according to the arguments passed to it. To know the details of the function, run

```
print range.__doc__
```

in an IDE. In brief, range (1000) as used in the example above will create a list [0, 1, 2, ···, 999]. The function can only create a list of integers which is why to generate a time instant that is a floating point number, we multiply a number from the list with the time step 10 μs. The for loop will assign each item in the list to the variable specified and execute the block of statements within the for loop that is marked by the indent. It is also possible to generate a step in the range function along with explicit start and stop numbers. Therefore,

```
range(3, 10, 2)
```

will generate the list [3, 5, 7, 9] as the function is now range (start, stop, step). If step is not specified, it is assumed to be 1. Therefore, range (3, 6) will produce [3, 4, 5]. It is also possible to generate a decreasing list as range (10, 3, −2) to produce [10, 8, 6, 4]. It should be evident from these examples that the list will start with the starting number specified and will add numbers such that:

```
number + step <= stop
```

For a decreasing list, the condition will be:

```
number + step >= stop
```

A number will not appear if it violates the end condition. Due to this reason, a list might be empty. For example, range (4, 3) will produce an empty list [] as the step when not specified is 1 and the start number 4 violates the condition $4 + 1 <= 3$. On the other hand, range (4, 3, −1) will produce a list with a single element [4] as only the start number satisfies the condition $4 - 1 >= 3$.

The next example will use the range function above along with another in-built function len() to perform multiplication of matrices. To iterate through the elements of a vector, the following loop statement would work:

```
for c1 in range(len(vector_a)):
```

`len(vector_a)` would return the length of the list `vector_a`. So `len([1, 3, 5, 7])` would produce 4, and the `range(len([1, 3, 5, 7]))` would in turn be equal to range (4) which would be [0, 1, 2, 3]. Therefore, c1 would iterate through the indices of the list `vector_a` which is what we want. So:

```
for c1 in range(len(vector_a)):
        print vector_a[c1]
```

will produce:

1
3
5
7

A matrix or an array in Python is merely a collection of lists within a list. So, a 4×4 matrix like an identity matrix of size 4×4 will be [[1,0,0,0], [0,1,0,0], [0,0,1,0], [0,0,0,1]]. If we wish to access each element of this matrix a row at a time, the following embedded loops would work:

```
for c1 in range(len(matrix_b)):
        for c2 in range(len(matrix_b[c1])):
```

The outer loop iterates over the elements of the list which are also lists being the rows of the matrix. The inner loop iterates over the elements of the rows. The function `len(matrix_b)` returns the number of elements within the list `matrix_b` which are the rows while `len(matrix_b[c1])` returns the length of the rows which are the number of columns of the matrix. The elements of the matrix can then be accessed as `matrix_b[c1][c2]`. Let us now multiply `matrix_b` and `vector_a`.

```
vector_c = []
for c1 in range(len(matrix_b)):
        if len(matrix_b[c1])==len(matrix_a):
                element_sum = 0.0
                for c2 in range(len(matrix_b[c1])):
                        element_sum += matrix_b[c1][c2]*
                                matrix_a[c2]
                vector_c.append(element_sum)
        else:
                print ''Matrices are not compatible''
                del vector_c
                break
```

The result of the multiplication is `vector_c`. A few things new in this example. Before multiplying two matrices or a matrix and a vector, it is necessary to check if they are compatible. In this case, the number of columns of matrix_b must be equal to the number of elements of vector_a or the matrices are not compatible. Since a matrix is a collection of lists within a list, this check needs to be performed for each row of matrix_b to ensure there is no error. In case of an incompatibility, the error is printed out and the resultant vector_c is deleted as the existing elements are irrelevant. The break statement causes the execution to jump out of the for loop and end the program. The break statement causes the execution to exit the loop in which it is found. If the break statement was encountered in the inner loop of nested loops, the execution would jump to the outer loop but would continue with that loop which in turn would lead to the inner loop being iterated again. Therefore, break is a convenient way to stop the execution of a particular loop. The result vector_c starts as an empty list. This need not be the case. vector_c can be initialized to a list with four elements, and each element can be assigned its value rather than the latest value being calculated and appended to vector_c as shown in the example. Quite often the resultant matrices are created and initialized. The above example merely shows how a simple example can be realized using in-built list functions.

```
vector_c = [0.0, 0.0, 0.0, 0.0]
for c1 in range(len(matrix_b)):
        if len(matrix_b[c1])==len(matrix_a):
                for c2 in range(len(matrix_b[c1])):
                        vector_c[c1] += matrix_b[c1][c2]*
                                        matrix_a[c2]
        else:
                print ''Matrices are not compatible''
                del vector_c
                break
```

Matrix manipulations are the backbone of the circuit simulator and for detailed examples on different types of matrix operations, read the code in solver.py of the circuit simulator. Besides matrix multiplication, code can be found to perform matrix triangularization and matrix inversion and to solve matrix equations.

2.8 Examples of Python Code in the Simulator—Strings

In the circuit simulator, strings play a major role in the user interface. Every input from the user whether through the command line or through spreadsheets is first extracted in the form of strings and then converted into other objects. This section will provide some code examples with respect to the circuit simulator which will help readers write their own Python code.

The user interface will be described in detail in the next chapter. However, at this point, to make this section relevant, a brief description of the use of spreadsheets as user interface is provided. In this circuit simulator, most of the inputs from the user are obtained from spreadsheets which are saved as Comma Separated Value (.csv) files. This implies that circuit schematics, component parameters, simulation

Table 2.1 A single row of a circuit schematic spreadsheet

Columns	A	B	C	D	E
Row	wire	`Resistor_R1`	wire	`Inductor_L1`	wire

parameters, and control function descriptors are all .csv files. When a spreadsheet is saved as a .csv file, the result is a text file where each row of the spreadsheet is a separate line in the text file and the cells in a row are separated from each other by commas. Furthermore, the lines in the .csv file are terminated by newline characters. As an example, consider the following row of a spreadsheet (Table 2.1)

When saved as a .csv file, the above row will look like the line of text below:

```
''wire,Resistor_R1,wire,Inductor_L1,wire\n''
```

In order to read the lines from a spreadsheet saved as a .csv file, the following code will do so:

```
nw_layout = []
ckt_file = open(''sample_ckt.csv'', ''r'')
for line in ckt_file:
        nw_layout.append(line)
```

The `open` statement will open the file in read mode, and the contents of the file will be in the object `ckt_file`. This object `ckt_file` contains a complete set of in-built functions for dealing with functions, and the reader is encouraged to check out the member functions with a file object. The for loop in this case loops through every string in the object `ckt_file` that contains the circuit layout. A line in the .csv file is terminated by a newline character which can be either "\n" or "\r" as shown in the example above.

With the lines of the .csv file extracted as strings and stored as items of the list `nw_layout`, the next step is to split each item that corresponds to the rows of the spreadsheet. To do so, we use the function `split()` associated with every string in Python. For example, let us suppose the row provided as an example above is the fifth row in `nw_layout`, i.e., `nw_layout[4]`. There is a difference in the list index and the row number as the row number starts with 1 while the list index starts with 0. Therefore,

```
nw_layout[4] = ''wire,Resistor_R1,wire,Inductor_L1,wire\n''
```

Performing,

```
nw_layout[4].split('',''')
```

which splits the string in `nw_layout[4]` with the comma "," as a delimiter and provides,

```
[''wire'', ''Resistor_R1'', ''wire'', ''Inductor_L1'', ''wire\n'']
```

The following block of code will extract an entire circuit layout from a .csv file and save it as a matrix, i.e., a list containing lists as described before.

```
nw_layout = []
ckt_file = open(''sample_ckt.csv'', ''r'')
for line in ckt_file:
        nw_layout.append(line.split('',''))
```

In this manner, the circuit in a spreadsheet saved as a .csv text file can be extracted into a matrix in Python. It is always possible that the user may leave leading or trailing white spaces in a cell. Also, it is essential to remove any newline characters "\n" or "\r" in the matrix that represents the circuit. This can be done by manipulating strings. Let us suppose, the element `nw_layout[0][1] = "Resistor_R1"` which has a leading white space. The following statement could eliminate this white space,

```
nw_layout[0][1] = nw_layout[0][1][1:]
```

The reason why this is possible is because a string is seen as a list (an array) of characters. So,

```
nw_layout[0][1] = '' Resistor_R1'' = ['' '', ''R'', ''e'',
                  ''s'', ''i'', ''s'', ''t'', ''o'',
                  ''r'', ''_'', ''R'', ''1'']
```

Python allows extracting elements from a list in the form of another list. Therefore [1:] is an extraction method where the element 1 to the last element (when not specified the second element is the last) is extracted to form another list and in this case a string. To give some more examples, consider,

```
nw_layout[0][1][1:4]
```

This means extract elements starting from 1 to 4-1. Therefore, the result will be "Res". The reader is encouraged to try out the above method with a few examples in an IDE to get the hang of it. In general for a list (or string) x of length N, the extraction:

```
x[n:m]
```

will produce a list of the elements of x starting from n to m-1. However, if `n>m`, the result is an empty list. The code block to remove leading white spaces is:

```
if nw_layout[0][1]:
        while nw_layout[0][1][0]=='' '':
                nw_layout[0][1] = nw_layout[0][1][1:]
                if not nw_layout[0][1]:
                        break
```

To begin with before attempting to make any change to any string, it is necessary to check if the string is not a null string. Attempting to access any element of a null string will produce an index out of range error. The next while loop will keep checking if the first element (character) of a string is a white space, and if so, it will remove it by extracting the string starting from character 1 to the last character. After every extraction, it is checked whether the resultant string is an empty string as in that case, we must exit the loop and not attempt to access any element of it.

In a similar manner as above, the trailing white spaces need to be removed as well. There are two methods to do this. For example, let us suppose, the above string was

```
nw_layout[0][1] = ''Resistor_R1  ''
```

To exclude the trailing white space, the following command will do:

```
nw_layout[0][1][:len(nw_layout[0][1])-1]
```

This will work because when the starting element is not specified it is assumed to be the 0th element of the string (or list) and the extraction will stop 1 element before the specified stop element which in this case is the length of the string. A more convenient way of performing the above task is:

```
nw_layout[0][1][:-1]
```

As stated before, the last element of a list (or a string) in Python can be accessed as -1th element. Therefore, element [-1] is equivalent to `[len(nw_layout[0][1])-1]` and therefore [:-1] will extract the string starting from the 0th element to second last element, thereby excluding the trailing white space. The code block for extracting the trailing white spaces would be as follows:

```
if nw_layout[0][1]:
        while nw_layout[0][1][-1]=='' '':
                nw_layout[0][1] = nw_layout[0][1][:-1]
                if not nw_layout[0][1]:
                        break
```

The newline characters "`\n`" or "`\r`" can be removed in a similar manner:

```
if nw_layout[0][1]:
        while nw_layout[0][1][-1]==''\n'' \
                or nw_layout[0][1][-1]==''\r'':
                nw_layout[0][1] = nw_layout[0][1][:-1]
                if not nw_layout[0][1]:
                        break
```

The newline characters "`\n`" or "`\r`" are only the last characters in a string as they terminate a row in the spreadsheet while saving it as a .csv file. The removal of leading and trailing white spaces and newline characters is performed for all data extracted from spreadsheets. This is because when two strings are compared, the addition of a single character like a white space is all it takes to break the algorithm.

2.9 Examples of Python Code in the Simulator—Dictionaries

Dictionaries are widely used in the circuit simulator as they allow fast access to data provided each data item is tagged with a unique key. In some applications, the key is unique while in some a unique key needs to be generated from the state of the circuit, and while accessing the data, a certain amount of bookkeeping is necessary.

The first example of a dictionary in the circuit simulator is for the type of components found in a circuit schematic. As an example, let us consider a circuit that has the following components:

```
Resistor_R1 , Inductor_L1 , Resistor_R2 , Capacitor_C1 , Ammeter_A1
```

In order to determine the type of a component, the string containing the entire component is split using the underscore (_) as follows:

```
ckt_comp.split(''_'')
```

If the component is `"Resistor_R1"`, the result of the split will be the list ["Resistor", "R1"]. Therefore, the component has been split into the type and the name of the component. Initially, a blank dictionary by the name of `components_found` is created.

```
components_found = {}
```

As a component is split into type and name, it is checked whether the type exists in `compon-ents_found`. This can be determined by checking for the keys in the component by:

```
if not ckt_comp.split(''_'')[0] in \
        components_found.keys():
```

Since `components_found` is an empty dictionary, `components_found.keys()` will return an empty list. Since the above condition is true, a key called "Resistor" can be created as follows:

```
components_found[ckt_comp.split(''_'')[0]] = []
```

which translates to:

```
components_found[''Resistor''] = []
```

We initialize the object corresponding to the item with key "Resistor" with an empty list as all the resistors found in the circuit will be added to this list. We then add the first resistor found as follows:

```
components_found[ckt_comp.split(''_'')[0]].append( \
                [ckt_comp.split(''_'')[1], comp_cell_position])
```

which translates to the following statement:

```
components_found[''Resistor''].append([''R1'', ''5E''])
```

A list containing the name of the component as the 0th item and the spreadsheet cell position as the 1st item are added. The spreadsheet cell position comp_cell_position of "5E" is completely arbitrary in the above example. This cell position is computed from the spreadsheet by a separate function. The dictionary components_found will now look like:

```
{''Resistor'': [[''R1'', ''5E'']]}
```

When Resistor_R2 needs to be added, the code will look like:

```
if not ckt_comp.split(''_'')[0] in components_found.keys():
        components_found[ckt_comp.split(''_'')[0]] = []
else:
        components_found[ckt_comp.split(''_'')[0]].append\
                ([ckt_comp.split(''_'')[1], \
                comp_cell_position])
```

In the above case, an else condition adds the list with Resistor_R2 since the dictionary key Resistor exists. After component Resistor_R2 has been added, components_found will look like:

```
{''Resistor'': [[''R1'', ''5E''], [''R2'', ''9D'']]}
```

In this manner, all resistors in the circuit will be added as lists with two elements (name and spreadsheet position) to the list corresponding to the key "Resistor".

When all the components in the example are added, the dictionary components_found will be:

```
{''Resistor'': [[''R1'', ''5E''], [''R2'', ''9D'']],
        ''Inductor'': [[''L1'', ''12F'']],
        ''Capacitor'': [[''C1'', ''4D'']],
        ''Ammeter'': [[''A1'', ''2C'']]}
```

The advantage of using dictionaries is that data corresponding to a key can be extracted much faster as Python has fast search functions. Without dictionaries, we would have to create our own data structures and generate search algorithms to extract data. For a large circuit with a large number of components of different types, by arranging them as dictionaries above, the parameter file for the components can be generated much faster rather than searching for the components in a general list. In this case, the key was the component type which is a predefined set of possibilities. However, the circuit simulator also uses dictionaries for cases where the keys are generated from the state of the circuit. For example, a circuit stores data related to the circuit based on the number of branches present in a certain closed loop. Suppose, there are six branches in the circuit. And also suppose that in a given closed loop only branches 2, 4, and 5 (counting from 0) are present. Therefore, a string can be used to represent this loop in terms of the branches present in it as:

```
''001011''
```

where "0" indicates that a branch is not present in the loop, while a "1" indicates that a branch is present in the loop. The above string can be generated from examining the loop for the presence of branches. The dictionary item for the above loop could be:

```
{''001011'': <Loop object>}
```

`<Loop object>` is fairly complex and is a collection of matrices and lists that generate all the necessary equations for the particular loop. In this manner, for every loop, a search tag can be generated and this can be used to look for existing keys in the dictionary. If the tag exists, the data are extracted. If the tag does not exist, the matrices are computed and a new item with that key tag is created.

2.10 Conclusions

This chapter has provided an overview of Python constructs used in the circuit simulator. A significant feature of Python is that everything is an object. Therefore, every data type contains in-built functions. An example being the `split` function available with strings which is extensively used in the simulator. One of the main features of Python has been the simplicity with which a complex data object can be constructed. An example has been provided of a list which contains floating point number, strings, and even other lists. The circuit simulator uses this functionality of Python extensively. As will be shown in the coming chapters, circuit analysis requires collecting information about the circuit in various forms—branch maps, loop maps, etc. Furthermore, with every iteration of the circuit analysis, updates to the circuit information result in updates to various data structures that decide how the next iteration of circuit analysis should proceed. All these need complex data structures which are a combination of lists, floating point numbers, strings, and sometimes user-defined objects. Using Python has therefore significantly reduced the time for developing algorithms.

Most of the code for the circuit simulator was developed in an evolutionary manner over time and with frequent modifications. Using Python and the in-built libraries made it easier to continuously modify algorithms and associated data structures. The main objective of this project has been to develop a circuit simulator that can simulate very large circuits without significant computational burden. This statement has a few contradictions. To begin with, by using a language as high level as Python, an inherent significant computational burden has been incorporated into the simulator. The circuit simulator still has the advantage of being lightweight as without a graphical user interface, in terms of computational burden, it is not too heavy on a processor. As an example, the circuit simulator has been used to run multiple simulations in parallel. To elaborate, let us assume a circuit needs to be simulated but several cases need to be run—different loads, different input conditions, different control parameters. If some or all of these cases can be run in parallel, it would save a significant amount of development time. As of now, with some of the systems simulated, multiple simultaneous simulations have been possible and this is a significant advantage of the circuit simulator. Therefore, choosing a high-level language such as Python for the circuit simulator has not had a significant impact on performance in terms of speed of simulations.

As stated before, one of the reasons for choosing Python has been the scientific and mathematical applications being developed in Python. At the time of writing this book, only in-built modules available with Python such as the `math` module have been used. However, using modules such as NumPy or SciPy are planned for the future. When simulating circuits, it is also necessary to perform post-processing on data such as Fourier analysis, wavelet transforms, and other signal processing functions. Software is being developed independently that performs these functions, and at a later date, the circuit simulator could be combined with these modules. This would be essential if the circuit simulator has to be comparable to commercial simulation software that offers a wide variety of processing and mathematical functions along with circuit simulation.

The next chapters will describe how the circuit simulator can be used to simulate power electronic circuits. The focus of these chapters will be to serve as a user manual for anyone wishing to use the circuit simulator. Chapter 3 will describe how circuit schematics can be developed and the library of components available with the simulator. Chapter 4 will describe how a control function can be written with this simulator and how special variables can be used to perform mathematical computations. Chapter 5 will describe how a simulation can be developed in stages with a shunt-connected VAR compensator as an example. These three chapters will contain code samples and will describe the importance of the features of the circuit simulator with examples. The brief Python tutorial provided in this chapter would enable a reader to understand the sample code while reading the following chapters. The reader could also read the blog which contains case studies with simulation results.

Chapter 3
User Interface

Abstract This chapter describes the interface that the simulator uses to interact with the user. The chapter describes the philosophy behind choosing spreadsheets as the mode of extracting information from the user whether to enter simulation parameters, circuit schematics, parameters of the components in the circuit schematics, and also the structure of control functions. The chapter describes how the structure of every component class in the simulator library and how the data entered by the user is processed by each component class. The chapter also describes the concept of how classes are instantiated for every component found resulting in objects and how these objects are referenced by the simulator. The chapter describes the execution flow in the simulator and how the simulator processes the data provided by the user and makes it available to the core simulation engine.

Keywords User interface · Spreadsheets · Circuit schematics · Object oriented programming · Classes · Instantiation · Pointers · Simulation execution flow

3.1 Introduction

Chapter 2 provided a basic introduction to Python and some code examples as they are used in the circuit simulator. With this chapter, a detailed explanation of the circuit simulator will begin. The user interface forms the outermost layer of the simulator as this is where the exchange of data with the user takes place. The user interface at this moment is a combination of spreadsheets and command line inputs since the simulator does not have a graphical user interface. Chapter 5 provides a detailed example of how to simulate a circuit and describes the step-by-step process of using the simulator. In this chapter, the focus will be on the different inputs the user can provide and how these are processed by the simulator.

The user interface is a combination of questions posed by the simulator and data in spreadsheets saved as .csv files. When the simulator needs data to be entered or updated in a spreadsheet, the simulator will use the command line or shell to

S. V. Iyer, *Simulating Nonlinear Circuits with Python Power Electronics*,
https://doi.org/10.1007/978-3-319-73984-7_3

inform the user about the spreadsheet that needs to be updated and to continue when ready. Basic error checking is provided at every stage to ensure that certain illegal operations are not performed, and when errors appear, the simulator informs the user about the whereabouts of the error in the spreadsheets. The user interface has been made as interactive as possible to ensure that the user will be informed about the input expected, corrections needed, and when the simulation results can be plotted.

At the time of writing this book, the user interface has been designed to be minimalistic for two reasons. First, to ensure that the simulator is lightweight and consumes the minimum possible resources from the processor. This is because the proposed application of the simulator is toward large circuits with several power converters. Second, by providing a minimal interface, the simulator is portable between operating systems and more importantly a simulation case that has been designed on one operating system can be directly used in another one. This makes it especially convenient when a user wishes to use an operating system to design a simulation case and then wishes to run the simulation for a long time on another computer through a virtual machine or remote login. In order to be able to use the simulator, a user needs Python, an editor for Python scripts, a spreadsheet software, and a third-party plotting software.

3.2 Circuit Representation

The first stage in the circuit simulator is to be able to generate a circuit schematic. A circuit schematic should be able to describe the components present in a circuit and also be able to describe how these components are connected to each other. The objective of the circuit simulator is to be able to simulate large circuits with several power converters. Therefore, the method chosen to represent circuits should be convenient for large circuits. However, at the same time, it would be desirable that the simulator be portable between operating systems. The user interface for providing data to the simulator is spreadsheets—whether it be the circuit schematic or the parameters of the components. One of the reasons for choosing spreadsheets as the mode of user input is that spreadsheets are ubiquitous and software for editing and creating them is available in every operating system. The circuit is "drawn" in a spreadsheet. Any spreadsheet software can be used as long as they can save the file as a Comma Separated Value (.csv) file, and each individual field is a string enclosed in quotes. A .csv file is essentially a text file which can be read by any file handling function in Python. In the .csv file, each row of the spreadsheet appears in a separate line and the contents of the cells in a row are separated by commas. At the same time, a .csv file can be opened in a spreadsheet software. Opening a .csv file in a spreadsheet software causes the user to lose some of the functions available with the software. However, as the sole aim of the .csv file is a mode of input to the simulator, the functions that are not available do not pose a disadvantage. The advantage of such representation is that a complex circuit can be described in a manner almost graphically but without a graphical user interface. The spreadsheet

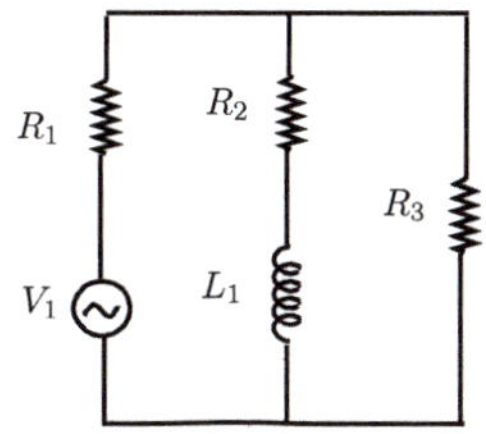

Fig. 3.1 Example of a circuit

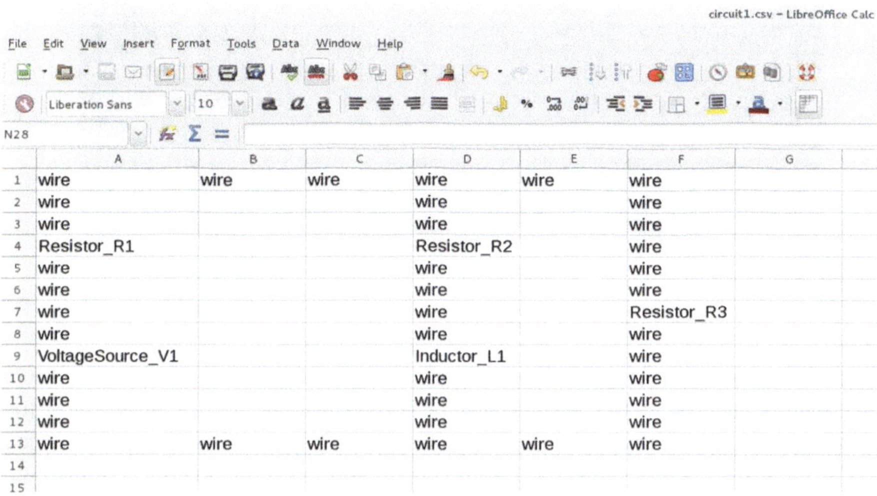

	A	B	C	D	E	F	G
1	wire	wire	wire	wire	wire	wire	
2	wire			wire		wire	
3	wire			wire		wire	
4	Resistor_R1			Resistor_R2		wire	
5	wire			wire		wire	
6	wire			wire		wire	
7	wire			wire		Resistor_R3	
8	wire			wire		wire	
9	VoltageSource_V1			Inductor_L1		wire	
10	wire			wire		wire	
11	wire			wire		wire	
12	wire			wire		wire	
13	wire	wire	wire	wire	wire	wire	
14							
15							

Fig. 3.2 Spreadsheet schematic of the circuit

can be treated as a two-dimensional grid where the circuit components are placed and are connected together by wires. Another advantage of such a two-dimensional grid is that a component placed in a cell can be referenced by the cell position which is unique. This section will describe how the circuit simulator processes a circuit schematic in a spreadsheet.

Consider the simple circuit in Fig. 3.1. Such a circuit can be described in a spreadsheet as follows in Fig. 3.2. As can be seen from the spreadsheet, a circuit can be described or "drawn" fairly conveniently as in this schematic, all that matters is the connectivity. The above schematic shows how components are connected together to form branches and those branches in turn are joined together. Most commercial software will provide a graphical user interface (GUI) where a user can actually use tools to draw a circuit. The components would have unique symbols for the components in the circuit shown above. In the spreadsheet, however, these symbols are replaced by keywords—VoltageSource, Resistor, Inductor. Usually, the GUI will also have an option where the components will have symbols and these symbols can be made visible in the schematic. In the spreadsheet, however, the symbols of

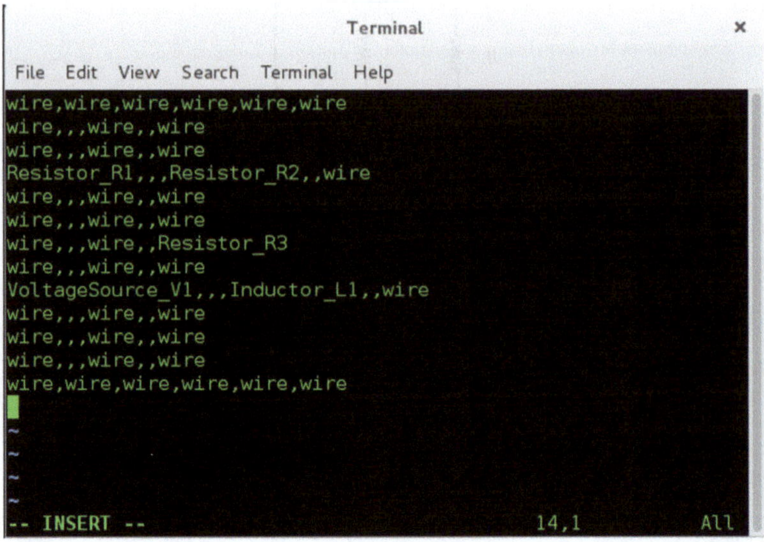

Fig. 3.3 Text file of the circuit schematic

the components—V1, R1, R2, R3, L1—are always present and are associated with the components. The "wire" elements in the spreadsheet are merely wire connectors used to join components together. Components can be placed next to each other if that is what the user prefers, but joining them with wires makes the circuit more presentable and also makes it easier to insert components in branches between existing components. Two components in adjacent cells imply an electrical connection. These could be two electrical components, two wire elements, or a wire element and a component. Conversely, an empty cell means no electrical connection. Therefore, if two components are separated by at least one empty cell, it means they are not physically connected along that path. The circuit has two nodes as can be seen from the figure and the spreadsheet. A node is simply formed by a wire element that has wire elements on at least three cells adjacent to it. A more detailed description of nodes, branches, and loops will be provided in the next chapter. This chapter will focus on how the user provides inputs to the circuit simulator.

When the spreadsheet is saved as a .csv file, it will appear in the manner shown in Fig. 3.3. This text file can be opened in any text editor and more importantly can be read and dealt with as an ordinary text file. The text file however is a convenient form of input to the circuit simulator. However, for purposes of editing the schematic file, the user can still open and edit the .csv file in any spreadsheet software like any other spreadsheet. Therefore, this provides a dual advantage—ease of reading in Python and also ease of use with respect to being a user interface. As stated before, when saving the spreadsheet as a .csv file, it is necessary to specify the delimiter between cells to be a comma. In the above text file, it can be seen that elements are separated by a comma. In the case of an empty cell, there will be no element between two commas.

However, this is still taken into consideration by the simulator as the absence of an element in a cell means no electrical connection and therefore is a null element. This will be explained in detail in the next section when the concept of the circuit matrix is introduced. Also, as stated before, each row of the spreadsheet will appear in a different row of the .csv file. This can however vary between software. Most software by default will add the newline character `"\n"` after each row. However, some will add `"\r"` between rows when saving them as .csv files. The simulator will look for both of these as newline characters. However, if some other newline character is used, it is advisable to look into the options and change it to `"\n"` or `"\r"`. Furthermore, the components in the .csv file above are strings. Therefore, when read by Python, `Resistor_R1` will be `"Resistor_R1"`. This is because the contents of each cell of the spreadsheet will be enclosed in quotes. Some spreadsheet software will use single quotes ‘ while some will use double quotes “. The simulator looks for either of these quotes. However, it is essential that some quotes either single or double are used so that the contents of the cells are read as strings. In this way, all the specialized functions available in Python to deal with strings can be used.

The next section will describe how the components are interpreted by the simulator and how each instance of an element forms a unique object.

3.3 Processing of Components

The previous section described how a circuit is represented in a spreadsheet. This section will describe how individual components within the circuit will be processed by the simulator. To begin with, there are three types of components in the circuit simulator as of now:

1. Components—Resistors, Capacitors, etc.
2. Jump labels
3. Wire elements

The components are the major part of the circuit simulator, and the later sections will be devoted completely to describing all the available components. Jump labels are merely connectors that connect branch segments together. The manner in which jump labels are dealt with is described in detail in Chap. 6 while describing how branches are determined. The wire elements mentioned in the previous section are connectors between components and help to add clarity to the circuit. Let us begin with components.

The example in the previous section had VoltageSource, Resistor, and Inductor. There was only one VoltageSource, `VoltageSource_V1`, and one Inductor, `Inductor_L1`, while there were multiple Resistor elements—`Resistor_R1`, `Resistor_R2`, and `Resistor_R3`. In most other simulation software, the circuit schematic will have a graphical user interface (GUI) where the components will have a unique symbol or shape, and upon opening their properties, a dialog box will open that lists their parameters. For example, the voltage source V1 would be the symbol of an ac or dc voltage source with the parameters being—the voltage peak, frequency, phase angle, dc offset. Another parameter which usually is listed but most users quite often do not notice is the name of the component. Simulators by default assign symbols using different naming schemes; for example, voltage sources could have automatic names of V1, V2, V3, etc. The user usually has the option to change these names and in larger circuits it is advisable to do so. In this circuit simulator, the name of the component has to be specified along with the type by the user. In `VoltageSource_V1`, `VoltageSource` is the type of the component and V1 is the name. The underscore (_) is the character that is used as a delimiter to separate the type of the component from the name of the component. Since the part before the underscore (_) is the type of the component, the user can only choose between a list of predefined components or create his own component class. The part after the underscore (_) is the name of the component. This is completely up to the user and can be any combination of letters and numbers. However, for any particular type of component, the name of every instance must be unique. Also, the underscore (_) is not permitted in any of these names as it is a delimiter between the type and name of the component.

Jump labels are mere connectors in the circuit schematic. As stated before, jump labels will be dealt with in detail in Chap. 6 but a brief description of their function will be provided here for completeness. Jump labels are used to connect two segments of a branch. Jump labels must occur in pairs. This is because a jump label must have a corresponding jump label from where the branch segment must continue. Therefore, the simulator while searching for a branch, on encountering a jump label, searches for the corresponding jump label and continues from that position onward in the schematic spreadsheet. Having only one jump label is an error as the simulator will not be able to complete a branch. More than two jump labels with the same name is an error as the simulator must have only one other position in the spreadsheet from where to continue the search for the branch. To define a jump label, a cell in a spreadsheet must have the first four letters "jump" and the rest is a unique identifier for the jump label. As an example, "jumpsourcevolta" consists of the jump identifier and the unique label "sourcevolta". It is advisable to be as verbose as possible while naming jump labels to make it easier to verify a circuit schematic.

The first stage of processing involves segregating the components found according to their type. Therefore, with respect to the circuit shown in the previous section, for the VoltageSource and Inductor component types there will be only one each namely V1 and L1, respectively. For the Resistor component, there will be three instances R1, R2, and R3. All the components in the circuit will be collected together in a dictionary called ComponentsFound. For the circuit shown and described, ComponentsFound will be:

```
{''VoltageSource'': [''V1'',''9A''], ''Inductor'': [''L1'',
                                                      ''9D''],
      ''Resistor'': [[''R1'',''4A''], [''R2'',''4D''],
                    [''R3'', ''7G'']]}
```

The keys of the dictionary ComponentsFound are the component types since they are unique, and the values for each key are the names of those components found in the circuit. Only the components found in a circuit will appear in the above dictionary. A sample code has been provided in Chap. 2, Sect. 2.9 to describe how this dictionary is created. When an element in the circuit is read and is split with the underscore (_) as the delimiter, the first part is the type of the component. If this component type exists in the dictionary ComponentsFound, the name of the component which is the second part is added to the existing list which is the value of the component key. If the component type does not exist, a new key is added to the dictionary ComponentsFound and the name of the component is added as the first in the list of components of that type found. The purpose of creating this dictionary is to check for two errors that a user could make but would cause the simulator to abort. The first error is if two components of the same type have the same name. For example, there cannot be two Resistor components with the name R1. This can be immediately spotted when collecting components together in a dictionary such as the one above as it can be checked if the name of a component that is being added to the value list already exists in that list and therefore is a duplicate component. The second error is that a component does not exist in the library. This could be due to a simple typing error or the user assumes that a component exists but is not yet in the basic library.

Once all the components have been added to the dictionary ComponentsFound and errors are eliminated, the next stage is to generate objects for each component found. An object in this case is different from the default Python construct. In Python, all variables and all functions are Python objects. This is because even if a variable contains merely a string, it has functions or methods that can operate on the string stored by the variable. When designing objects for the components in the simulator, each object contains data for the type of component it represents and contains methods or functions that operate on this data and transfers the data to the circuit simulator. Each component type has a different class with that name. Therefore, for resistors, there is a Resistor class and all the Resistor components found in the circuit will have an object that is an instance of this Resistor class. In this manner, each component found in the circuit will have a unique object. A detailed description of the data contained in the objects for each component type and also the type of methods that operate on this data will be provided in the next few sections. This section will address another concept. How these objects are named, referenced, and called by the circuit simulator. To elaborate on this, as time progresses in each iteration of the simulation, the circuit simulator will access the data in the object of each component, update the matrices used for loop and nodal analysis, and solve these matrix equations. Subsequent to the loop and nodal analysis, the circuit simulator will update the data in the object in each component. Therefore, the circuit simulator needs an efficient manner of accessing the object of each component and manipulating the component's data.

The first part is to automatically name the object for each component. This name should be unique. There are two choices in this matter. The first is the type of the component and the name exactly as it appears in the circuit schematic. For example, Resistor_R1 could have an object name of Resistor_R1. There could be several Resistor components in the schematic, but there cannot be two Resistor components with the name R1. Therefore, this name would be unique. The advantage is that names of the objects would be obvious and easy to debug as opposed to a random serial number. The disadvantage is that for a large circuit with several components, using this naming scheme might result in long names which might be inefficient when accessing the objects. The other option is the cell position. For example, in the above example, the cell position for Resistor_R1 is 4A. A cell in the spreadsheet cannot contain two components. Therefore, by using the cell position, a unique name for the object of each component is guaranteed. The advantage of this method is that string sizes for the object names will be much smaller. The disadvantage would be that the names would not be as obvious as the names in the circuit schematic which would make it necessary to cross-reference while debugging. As of the time of writing this book, the convention of naming the object of each component in the circuit is with respect to the cell position in the spreadsheet. Therefore, for the circuit chosen an example, the dictionary containing the objects of all the components in the circuits is:

```
{``4A'':<ObjectR1 >, ``9A'':<ObjectV1 >, ``4D'':<ObjectR2 >,
        ``9D'':<ObjectL1 >, ``7G'':<ObjectR3 >}
```

In the above dictionary, the objects denoted within $\langle\rangle$ are just for description of the concept. In reality, these will be pointers to the objects created when Python creates an instance of a class of a component type. Therefore, the value of each key will be an arbitrary value that a user will not understand as it is generated by the Python compiler.

With a background on how a user-defined circuit is processed by the circuit simulator, we will now examine how a component object is formed. The following section will describe the data contained by each object type and therefore will describe the parameters that can be specified for each object. The section after that will describe the functions or methods that act on this data and update the system matrices used by the circuit simulator.

3.4 Data Structures of Components

The data contained in component objects fall into three types— those that are automatically generated, those that are entered by the user, and those that are updated as the simulation progresses. To begin with, we will describe the data that are generated automatically and are more or less common for all types of components. The fundamental variables present in each class definition of a component type are Type, Position, Tag, and Index. Type is the type of the component object—for example, any object of component type resistor will have this as "Resistor". Position is the position of the component in the spreadsheet. However, this position will not be exactly the same as the spreadsheet position as that has both numbers and letters. Instead, the position will be the position of the component in the matrix which is extracted from the spreadsheet. Thus, resistor R1 which is at cell position 4A in the spreadsheet will be [3, 0] in the circuit matrix and this will be the value of Position.

Fig. 3.4 Resistor

Tag is the name of the component in the spreadsheet. So for Resistor_R1 at 4A, the tag will be "R1". Index is a default variable that is generated as the components are read from the circuit. This helps in debugging but has no relevance to the operation of the circuit simulator.

The next few variables that are common to all are status flags. These are "IsMeter", "IsControllable", and "HasVoltage", and their values are Boolean variables—True or False. The variable "IsMeter" is True if the component has the ability to measure a physical quantity. As of now, the simulator has only Voltmeter and Ammeter components as meters where this variable is set to True. With this variable, the simulator is able to determine the components whose measured values are to be stored in the output data file with which the user can plot waveforms. "IsControllable" is True if a component can be controlled and is False in all other cases. Component types that have this variable True are ControlledVoltageSource, VariableResistor, VariableInductor, and Switch. It is always possible to conceive of other elements that can have a controllable status in which case they will also have this variable set to True. "HasVoltage" is a variable that is True if the component has a voltage source. The component types VoltageSource and ControlledVoltageSource have this variable set to True as their function in the simulator is to be able to produce a voltage source. However, any component can have a voltage source that could serve as parasitic drop—for example, Diodes and Switches can have a voltage drop programmed besides the resistance. When a component having this variable "HasVoltage" to be True is found, the simulator adds the component to the SourceList. This SourceList is used to generate the input vector containing all the voltage sources which the circuit simulator will use to perform circuit analysis. This concept will be described in the later chapters when describing loop and nodal analysis.

Now we will begin by describing the distinct data contained in each component type. The description below will be of all the components that can be included in the circuit simulator.

3.4.1 Resistor

The most basic component type has the only distinct variable being the value of the resistance and this variable is Resistor. The default value of the variable when an object of this class is instantiated is 100 ohms. The symbol for the resistor is in Fig. 3.4. The list of variables for resistor component types is in Table 3.1.

Table 3.1 Resistor parameters

Variable	Default value
Resistor	100
Type	"Resistor"
Tag	From the circuit
Position	From the circuit
HasVoltage	False
IsMeter	False
IsControllable	False

Fig. 3.5 Inductor

Fig. 3.6 VoltageSource

3.4.2 Inductor

Another basic component type has the only distinct variable being the value of the inductor and this variable is Inductor. The default value of the variable when an object of this class is instantiated is 0.001 Henry. The symbol for the inductor is in Fig. 3.5. The list of variables for inductor component types is in Table 3.2.

3.4.3 VoltageSource

This is a basic component type that has a few more variables than the previous two. It is assumed that the voltage source is an ac source of a constant magnitude and frequency, with an initial phase angle advance or delay and with a possible dc offset.

Table 3.2 Inductor parameters

Variable	Default value
Inductor	0.001
Type	"Inductor"
Tag	From the circuit
Position	From the circuit
HasVoltage	False
IsMeter	False
IsControllable	False

Table 3.3 VoltageSource parameters

Variable	Default value
Peak	120
Frequency	60
Phase	0
DcOffset	0
VoltageOutput	0
Polarity	[–1, –1]
Type	"VoltageSource"
Tag	From the circuit
Position	From the circuit
HasVoltage	True
IsMeter	False
IsControllable	False

Therefore, these are the parameters of the component type. Besides this, a voltage source also has a polarity. When a voltage source is inserted in a branch, the positive polarity will be toward one of the nodes of the branch. This polarity is given the default value of [–1, –1] when the object is created.

However when the cell position of the VoltageSource is read, the polarity is arbitrarily assigned to the next element in the branch and is left to the user to modify it. The symbol of the VoltageSource is shown in Fig. 3.6. The parameters of the component type are listed in Table 3.3.

VoltageOutput is the final output of the source computed as:

```
Peak*sin(2*pi*Frequency*Time + Phase) + DcOffset
```

In this case, the component type has variable "HasVoltage" to be True and therefore any component of this type will be added to the SourceList by the simulator. If the user wants a dc source instead of an ac source, the peak can be made zero and the DcOffset variable will contain the magnitude of the voltage source. The ac component is by default calculated as a sinusoid with respect to the current time instant. However, the Phase variable allows the user to define a phase advance or delay between different voltage sources in the circuit.

3.4.4 Capacitor

This is the last of the basic passive components. The symbol for the capacitor is in Fig. 3.7. The capacitor in this circuit simulator has been modeled as a voltage source for the purpose of loop and nodal analysis. The voltage of the capacitor is updated based on the current flowing through it by integrating the differential equation:

Fig. 3.7 Capacitor

Table 3.4 Capacitor parameters

Variable	Default value
Capacitor	10.0e-6 (Farad)
Voltage	0
Polarity	[–1, –1]
Type	“Capacitor”
Tag	From the circuit
Position	From the circuit
HasVoltage	True
IsMeter	False
IsControllable	False

$$v = \frac{1}{C}\int idt \quad \rightarrow \quad v_n = v_{n-1} + \frac{1}{C} i_n \Delta T \tag{3.1}$$

The integration is performed in one of the methods that will be described in the next section. However, for the circuit simulator, in terms of loop or nodal analysis, the capacitor is as good as a voltage source. One of the variables of this component is the value and is Capacitor. The other variable is the polarity of the capacitor which is mainly important in the case of dc capacitors. The list of parameters is in Table 3.4.

3.4.5 Ammeter

This meter is used for measuring the current in the branch in which it is connected. It is one of the most fundamental parts of any circuit as the current through different branches is something any circuit designer would like to measure and record. The current measured by the Ammeter is automatically written in the output file specified by the user. Physically, an Ammeter has the lowest possible resistance so that it does not alter the current in a branch in which it is connected. In simulation, this lowest possible resistance can be made zero and it is. The only parameter the Ammeter has is the polarity—current flowing toward which node in the branch is considered positive? The symbol of the Ammeter is in Fig. 3.8. By having a zero resistance, an Ammeter does not affect a branch and therefore the resulting loop and nodal analysis in any way. As will be explained in the later chapters, the branch currents are updated at the end of every simulation iteration following loop and nodal analysis. Depending on the branch an Ammeter is present in, the current of the Ammeter is equated to the

Fig. 3.8 Ammeter

Table 3.5 Ammeter parameters

Variable	Default value
CurrentMeasured	0
Polarity	[−1, −1]
Type	"Ammeter"
Tag	From the circuit
Position	From the circuit
HasVoltage	False
IsMeter	True
IsControllable	False

branch current at the end of every simulation iteration and this current is recorded in the output data file. The list of parameters is in Table 3.5.

3.4.6 Voltmeter

This meter is used for measuring the voltage between a pair of nodes by connecting a branch with a Voltmeter across these pair of nodes. Just like the Ammeter, it is an extremely critical part of the simulator for measuring the voltage across different parts of the circuit. Unlike the Ammeter before, the Voltmeter is a bit complicated. Physically, a Voltmeter should have a resistance large enough such that the current drawn by the Voltmeter is negligible and does not affect the currents in the actual circuit. Ideally, an infinite resistance would ensure that the circuit is not disturbed by the connection of the Voltmeter. However, an infinite resistance is not possible with respect to the computations involved in the loop and nodal analysis. Therefore, the resistance of the Voltmeter is chosen to be "very large". The concept of "very large" is relative and needs some explanation. There are circuits that are inherently low power—for example, a dc–dc converter of a cellular phone would have a maximum current which might be less than an ampere for most operating conditions. On the other hand, most of the other circuits would have their normal operating currents to be in the range of a few amperes. The circuit simulator is focused on medium to high power circuits with typical applications being in grid-connected renewable energy sources. Therefore, it has been assumed that a negligible current for an element to draw would be 1 μA and this assumption has been applied to the Voltmeter too. The resistance of the Voltmeter is calculated such that it will draw a current not more than 1 μA. The only other parameter that remains is that maximum possible voltage that could appear across the Voltmeter. This parameter need only be an approximation

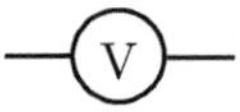

Fig. 3.9 Voltmeter

Table 3.6 Voltmeter parameters

Variable	Default value
VoltageLevel	120
Polarity	[–1, –1]
VoltageMeasured	0
Type	"Voltmeter"
Tag	From the circuit
Position	From the circuit
HasVoltage	False
IsMeter	True
IsControllable	False

and could be twice or thrice the magnitude of the voltage sources in the system. The symbol of the Voltmeter is in Fig. 3.9, and the parameters are in Table 3.6.

3.4.7 *VariableResistor*

This is an extension of the Resistor class before with the value of the resistor being controllable by user-written code. The symbol for the VariableResistor is in Fig. 3.10, while the parameters are listed in Table 3.7. The value of the VariableResistor is still called by the variable Resistor. The VariableResistor has a variable called ControlTags. All controllable components have this variable. ControlTags is a list of all the control variables that the component type has. In this case, there is only one controllable variable which has a default name "Resistance". This controllable variable is how the circuit simulator links the resistor of the VariableResistor to the control code written by the user. This controllable variable should therefore be changed to a name that is unique and preferably relevant to the function of the VariableResistor so that the user may always identify in the control code, which VariableResistor's value is being changed. ControlValues are the list of values of the controllable variables. In this case, there is only one value as there is only one variable. However, in the case of multiple controlled variables, the elements in the ControlValues list will have a one-to-one correspondence with the names in ControlTags. The concept of how control code can be written by the user and how these are linked to the component parameters will be described in Chap. 4. Finally, the IsControllable variable in this component type now has a value of True.

Fig. 3.10 VariableResistor

Table 3.7 VariableResistor parameters

Variable	Default value
Resistor	100
ControlTags	["Resistance"]
ControlValues	[100]
Type	"VariableResistor"
Tag	From the circuit
Position	From the circuit
HasVoltage	False
IsMeter	False
IsControllable	True

Fig. 3.11 VariableInductor

Table 3.8 VariableInductor parameters

Variable	Default value
Inductor	0.001
ControlTags	["Inductance"]
ControlValues	[0.001]
Type	"VariableInductor"
Tag	From the circuit
Position	From the circuit
HasVoltage	False
IsMeter	False
IsControllable	True

3.4.8 VariableInductor

This is an extension of the Inductor class before with the value of the inductor being controllable by user-written code. The symbol for VariableInductor is in Fig. 3.11. The variables are similar to the VariableResistor component type described before and are listed in Table 3.8. The controllable variable is called "Inductance" by default, and the value of the VariableInductor can be accessed through this variable in the control code. This controllable variable can be changed by the user, and it is recommended to choose a unique and characteristic name for this variable.

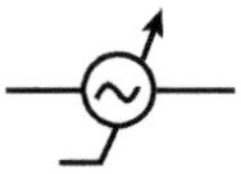

Fig. 3.12 ControlledVoltage Source

Table 3.9 ControlledVoltageSource parameters

Variable	Default value
ControlTags	["VoltageOutput"]
ControlValues	[0.0]
Polarity	[–1, –1]
Type	"ControlledVoltageSource"
Tag	From the circuit
Position	From the circuit
HasVoltage	True
IsMeter	False
IsControllable	True

3.4.9 ControlledVoltageSource

This is a modification of the VoltageSource class before with the value of the VoltageOutput being controllable by user-written code. The symbol for ControlledVoltageSource is in Fig. 3.12. The controllable variable is "VoltageOutput". This can be directly changed by the user control code in any manner decided by the user. The list of parameters is in Table 3.9.

3.4.10 Diode

The symbol of the Diode is in Fig. 3.13. The parameters of the Diode are listed in Table 3.10. The Diode is a nonlinear element that can either be in a low resistance ON state or a high resistance OFF state. The Diode has a polarity and can only conduct current flowing from the anode to the cathode, and the user specifies which direction the cathode is in. When the Diode is forward biased which means the voltage difference between the anode and the cathode is greater than a threshold which is the barrier voltage, the Diode turns ON and enters the low resistance ResistanceOn state. When the current through this Diode in the ON state becomes negative, the Diode turns OFF and enters the high resistance ResistanceOff state when it conducts a negligible current. As with the Voltmeter, the resistance of the Diode in the OFF state should be so high that the current it draws is negligible and does not disturb the rest of the circuit. Therefore, the Diode also has a VoltageLevel variable to provide an estimate of the maximum possible voltage the Diode could face across it in the reverse biased state. With this voltage, the Off resistance of the Diode ResistanceOff should

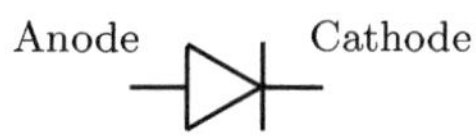

Fig. 3.13 Diode

Table 3.10 Diode parameters

Variable	Default value
VoltageLevel	120
Resistor	ResistanceOff
Polarity	[–1, –1]
Status	Off
Voltage	0
Current	0
Type	"Diode"
Tag	From the circuit
Position	From the circuit
HasVoltage	False
IsMeter	False
IsControllable	False

Fig. 3.14 Switch

be such that it will draw a current of not more than 1 μA. As with the Voltmeter, this VoltageLevel need only be an estimate and a multiplier can be used with the system voltage to arrive at this number. The variable status indicates the state of the Diode—whether it is in the ON or OFF state.

3.4.11 Switch

The symbol for the Switch is in Fig. 3.14. The parameters of a Switch are in Table 3.11. The Switch is similar to the Diode in the sense that it allows current to flow in only one direction—from anode to cathode. Also, similar to the Diode, the Switch has two modes of operation—low resistance (ResistanceOn) mode of operation and high resistance (ResistanceOff) mode of operation. Unlike the Diode, the Switch has a control signal which is the gating signal. Only when the gating signal is high (1) will the Switch turn on when it is forward biased. Conversely, when the gating signal goes low (0), the Switch will turn off even if it is conducting a current. As with the previous controllable component types, the Switch has the ControlTags variable which contains a default variable "Control". This control variable should be changed by the user to be unique and characteristic of the Switch. The control variable is accessible through control code written by the user.

Table 3.11 Switch parameters

Variable	Default value
VoltageLevel	120
Resistor	ResistorOff
Polarity	[–1, –1]
ControlTags	["Control"]
ControlValues	[0]
Status	Off
Voltage	0
Current	0
Type	"Switch"
Tag	From the circuit
Position	From the circuit
HasVoltage	False
IsMeter	False
IsControllable	True

3.5 Logical Flow of the Simulation

With the description of the data contained by the component objects, we will now examine how the circuit simulator extracts data from these component objects for circuit analysis and updates the data at the end of every iteration. The techniques of circuit analysis namely loop analysis and nodal analysis will be described in the later chapters. In the following description, the circuit analysis has been considered as a black box with the focus being the exchange of data between the core simulation engine and the component objects.

To begin with, let us examine how the simulator starts up. The flowchart in Fig. 3.15 shows the start up routine, and a description of every step is as follows.

3.5.1 Launch Simulator

At the time of writing this book, this needs to be done by executing the main file "circuit_solver.py" using Python. A number of Python environments exist where you can execute programs besides also having an interactive shell. As an example, in a Unix environment, using a shell such as BASH, the following command will launch the circuit simulator:

```
$ python circuit_solver.py
```

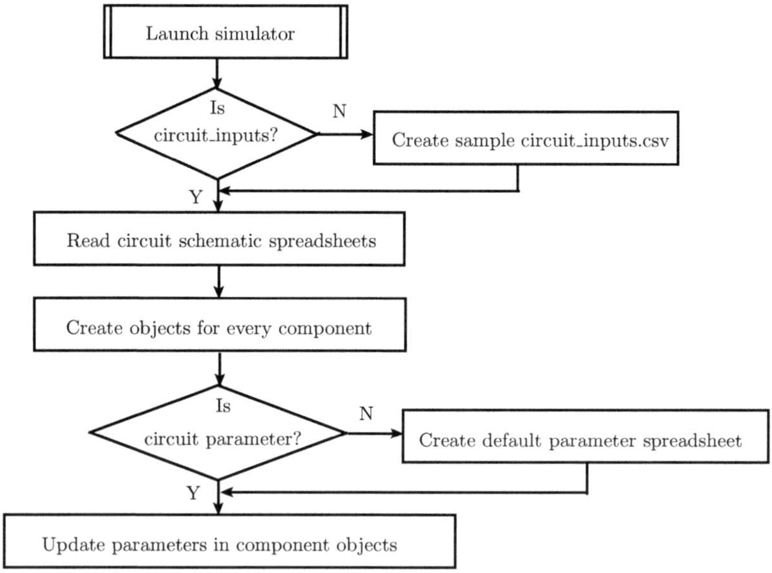

Fig. 3.15 Simulation process

3.5.2 Simulation Parameters

After launching the circuit simulator, the simulator looks for parameters in the circuit_inputs.csv spreadsheet. This spreadsheet has the data in Table 3.12 and is similar to the simulation parameters dialog box in most commercial software. The circuit file has been chosen for the example chosen in the beginning of this chapter. The user is encouraged to read the guidelines of the operating system being used to know what are the legal file names and to follow these while naming circuit files. The only requirement from the circuit simulator is that it should be a Comma Separated Value (.csv) file. The time duration and simulation time step are fairly self explanatory—time duration is the maximum time limit of simulation, and time step is the time increment after every iteration. Time step for data storage is an entry provided in case the user does not wish to record data at every simulation instant as by skipping data points the size of the output file will decrease and will be easier to plot. The name of the output file is again up to the user, and it is recommended that the file be consistent with guidelines from the operating system.

For simulating the circuit1.csv example, we do not have controllable components and at this point we do not wish to do data processing. Therefore, control files have been left blank. It should be noted however that it is not necessary to have controllable components in order to have a control file. A control file in this case could perform processing such as calculating power, harmonics, peak values. Control files will be dealt with in detail in Chap. 4. The simulator offers the option to split the output file. This is particularly useful for simulations of long duration as the output data

Table 3.12 Simulation Parameters

Name of the circuit file	circuit1.csv
Time duration of simulation	180 (s)
Time step of simulation	10.0e-6 (s)
Time step of data storage	10.0e-6 (s)
Name of data storage file	circuit_output.dat
Name of control files	
Split the output file?	Yes
Length of time windows	30 (s)

files can be very large in size—in hundreds of megabytes or even gigabytes. When plotting files that are large, most plotting software tends to stall or hang. It is usually advisable to split the files and plot the intervals that are needed. If an interval is of no importance, the data files corresponding to that interval can be deleted or alternatively compressed to save disk space.

On launching the circuit simulator, the simulator will look for a circuit_inputs.csv spreadsheet. In case a new project has been started and such a spreadsheet has not been found, the simulator will create a sample circuit_inputs.csv spreadsheet with default values and ask the user to change it according to the user's requirements. If a circuit_inputs.csv has been found which would be the case if the simulation is being repeated, the simulator asks the user to verify the simulation parameters.

The circuit schematic file specified in the table above is read by the simulator. There are two aspects to processing the circuit schematic. The first step is the determination of the components in the circuit which has been described in this chapter. The second step is to determine the connectivity between the components in the circuit. This topic will be covered in Chap. 6. In Chap. 6, it is described how the simulator extracts the circuit layout and stores it as a matrix. An element of the matrix can be a component, a wire, a jump label, or can be empty. However, as will be shown in Chap. 6, this circuit matrix is used primarily for determining the nodes, branches, and loops in the circuit in order to perform circuit analysis. This function has not been described in this chapter, as the user does not play an active role in the determination of these nodes, branches, and loops, while this chapter describes the interaction between the user and the simulator.

3.5.3 Create Component Objects

The circuit simulator having extracted the components in the previous step will create objects for each component found. This is where the object-oriented nature of Python is exploited. In object-oriented programming, data and code go hand in hand. In traditional coding, code acts on data. However, as the nature of data varies, the code will need to be designed so as to be able to function on every type of data. As an example, in the circuit simulator, each type of component has different types of data. Therefore, if a single set of functions containing code were to be designed for all the components present in a circuit, these functions would be fairly cumbersome. Instead, by defining each component type as a class with data and associated functions containing code, a far more efficient method of programming is achieved.

This chapter has described the data present in each type of component and the function of these data items. As shown, some of the data are automatically generated by the simulator while some are modified by the user. Based on the type of a component, the simulator will instantiate the corresponding class creating an object that will be referenced by the cell position of the component. The data in each component will be the default values when the component is instantiated. This is to ensure that the user has a reference with which to modify values. Each component object will also contain member functions that transfer data between the simulator circuit analysis functions and the component. Moreover, some member functions will process component data as many of the components have outputs.

3.5.4 Circuit Parameters File

When the user creates a new circuit schematic, for example in this case circuit1.csv, the simulator automatically creates objects for all the components. As stated before, some of the component data are automatically generated by the simulator while some are assigned default values which can be changed by the user. As an example, the user cannot change the name (tag) of a component or the schematic spreadsheet cell position of the component. These are generated by the simulator, and if the user wishes to change these, the change has to be made in the circuit schematic spreadsheet. As for parameters which can be changed by the user, the user can modify them by making changes to a parameter spreadsheet.

For a circuit schematic circuit1.csv, the parameter file has the name circuit1_params.csv. Essentially for any circuit file x.csv, the parameter spreadsheet is called x_params.csv. The user is now free to change the parameters of each component according to the user's design specifications. In case the simulation is being repeated, this parameter spreadsheet already exists and has been altered by the user. Therefore, the simulator merely asks the user to verify the parameters in the spreadsheet. A detailed description of a simulation example is provided in Chap. 5 where detailed description of circuit parameters is provided.

3.5.5 Update Component Parameters

After the user changes and confirms the parameters of the circuit in the parameter spreadsheet, the simulator reads these parameters and updates the parameters in each component object. This chapter does not describe in detail the process of extracting information from the spreadsheet. However, these spreadsheets need to be saved as .csv files and therefore are available to the simulator as text files. The reader can refer to Chap. 2 on how a row of a .csv file is extracted by the simulator.

These are the steps in the simulation that require inputs from the user. Now that the simulator has the updated circuit schematic and parameters, the simulator will generate information about the circuit in terms of the nodes, branches, and loops of the circuit. This procedure is described in Chap. 6. The subsequent chapters will describe how circuits are simulated. However, before concluding this chapter, we will look at how information is exchanged between component objects and the simulator during the simulation process.

3.6 Iterative Procedure During the Simulation

The flowchart in Fig. 3.16 shows the iterative process of the simulation. Most of the details of each step are contained in the following chapters. However, the description in this section is to enable the user to visualize how the simulator uses component objects.

Fig. 3.16 Iterative process

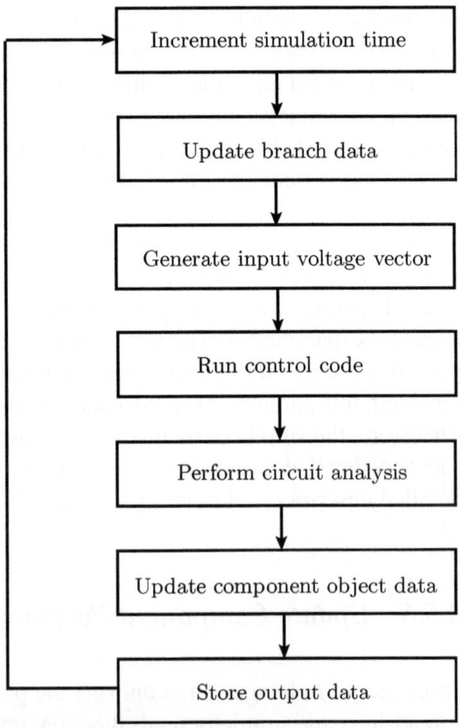

3.6.1 Simulation Time Instant

The simulation time instant is the running time of the simulation. It is updated at every iteration with the simulation time step specified in the simulation parameter spreadsheet. This simulation time instant is quite often used for generating the input voltage vector used for performing circuit analysis as will be described in Chap. 7. For most simulations, this update is fairly simple. The time instant update however gets complicated when there is a control file that has a different update time interval than the simulation time step. Details on this scheduling will be described in Chap. 4 when control is dealt with. In the presence of control functions, it is possible that the time step of a control function is completely different from the simulation time step. In that case, the time scheduler of the simulator ensures that a control function executes at the exact intended time instant.

3.6.2 Update Branch Data

The simulator processes circuit data in two stages. The first stage is to determine how the parameters of the circuit components affect the parameters of the branches in the circuit. As will be shown in Chaps. 7 and 8, the branches of the circuit are used in performing circuit analysis. At the end of a simulation iteration, there could have been changes in the circuit due to updated loop and branch currents. For example, a Switch could have turned ON, a Diode could have turned off, or a VariableResistor could have changed its value due to control code. These will produce changes in the parameters of the component objects, and these changes in parameters need to be fed to the circuit simulator. The manner of doing so is to identify the branch in which a component appears in and update the information of that branch. This information is the resistance of the branch, inductance of the branch, and the voltage sources that appear in the branch.

3.6.3 Generate Input Voltage

Without going into the details of the circuit analysis which is described in the later chapters, the input to the simulator is the voltage sources in the circuit. All components that have a voltage source as described in the parameters in the previous section will appear in a separate list called SourceList. At every iteration, the voltages in the source list will be updated from the voltages in the component objects. For example, consider the VoltageSource_V1 which is a simple sinusoid of peak 100 V and frequency 60 Hz in the example circuit1.csv. At every time instant, the voltage output of this VoltageSource_V1 will be updated in the corresponding component object and this will be transferred to the corresponding value in the list SourceList. In this manner, the input vector to the circuit analysis is updated and this in turn will be used to solve equations that produce the branch currents and node voltages. Details on this are in Chaps. 7 and 8.

3.6.4 Run Control Code

As shown in the spreadsheet circuit_inputs.csv, the simulator parameters have a separate row for the control functions that the user wishes to use. It should be reiterated at this point that a control function need not directly or indirectly perform a control action and a circuit does not need a controllable component in order to process control functions. Control functions could perform data or signal processing. Details of how a simulator processes a control function are provided in Chap. 4. Furthermore, Chap. 5 provides a detailed example of a simulation case study with multiple control functions and how these are interfaced. The simulator runs every control code provided by the user at every simulation time instant. As will be described in Chap. 4, a control function can have several different types of variables. Some of these variables are mere control signals while some can affect the state of the circuit. These updates could be resistances, inductances, or voltage outputs of controllable components.

3.6.5 Perform Circuit Analysis

The simulator performs circuit analysis as a combination of loop analysis and nodal analysis. Chapter 7 is dedicated to loop analysis, while Chap. 8 is dedicated to nodal analysis. Loop analysis has the advantage of resulting in smaller system matrices while solving ordinary differential equations. This is particularly the case when circuits become larger and more complex. In a circuit with N nodes and B branches, the number of independent loops is $B-N+1$. As an example, a circuit with 111 branches and 74 nodes will have 38 independent loops. Therefore, the number of equations to be solved will be 38. In contrast, nodal analysis will solve matrices by the size of $(N-1) \times (N-1)$ which is 73×73. Nodal analysis is used by the circuit simulator; however, as will be described in Chap. 8, the purpose of nodal analysis is to solve the stiff part of the circuit and to determine conductivity of nonlinear devices. Therefore, by limiting the use of nodal analysis, the computational burden of the simulator is decreased, an advantage that is expected to ease the simulation of larger circuits.

3.6.6 *Update Component Objects*

Subsequent to the circuit analysis, the loop currents, branch currents, and node voltages of the circuit will be updated. These changes in the currents and voltages of the circuits will result in changes in the circuit components. As a simple example, for components like Ammeter and Voltmeter, the measured currents and voltage drops will be updated with the branch currents and node voltages. For some other components, particularly nonlinear components like Diodes and Switches, the currents and voltages could cause them to change their conductivity. For example, if the branch current changes such that a conducting Diode present in the branch detects a negative current, the Diode will turn off. Therefore, after performing circuit analysis, it becomes necessary to use branch data to update the data in component objects. From a computational perspective, by updating component data at the end of a simulation iteration, the associated member functions of a component will always have available to them the latest data. This ensures that every object of the simulator is up-to-date.

3.6.7 *Write Output Data*

The output data file allows the user to plot waveforms to check the progress of the simulation. The simulator offers a few options that expand the usefulness of the simulator. In terms of plotting data, besides the Ammeters and Voltmeters that are by default written to the output data file, the user can also plot certain variables in control functions. This makes it easier for the user to debug control code or to plot waveforms of quantities like power, peak voltage. The simulator asks the user in circuit_inputs.csv to specify the name of the output data file. Another feature provided is specially targeted to simulation of large circuits. One of the main challenges of simulation large circuits is the plotting of waveforms as the data needed to be plotted may be very large. A common method used is to plot only the latest data generated by the simulation. However, in many cases, this is unacceptable as the user needs waveforms of every transient including the start-up of the circuit. The simulator allows a user to split the data files. By specifying "Yes" to the question "Split the output file?", and specifying a time window of every split, the user can break up the output files into a number of files. Each output file will have the serial number appended to it. The advantage of doing so is that the user can plot as many output data files as needed while ignoring those that are not needed or simply deleting some if hard disk space is a constraint. In the future, an application to compress these data files could also be provided as these output data files are mere text files and compressing them could save a significant amount of disk space.

3.7 Conclusions

This chapter has described how the user interacts with the simulator and how this user interface impacts the simulator. This chapter in particular focuses on the structure of component data and the parameters that the user is free to modify. The chapter also describes the stages of the simulation process, with each stage being described in detail in subsequent chapters. This chapter has not described how a user can include control functions into the simulator as the entire Chap. 4 is devoted to it. Moreover, Chap. 5 describes an entire simulation case study and the reader is strongly encouraged to read the chapter.

The user interface of the simulator has been designed so as to provide most of the features of a commercial simulation software. Commercial software allows the user to change the parameters of the simulation with respect to time duration, simulation time step, and some other options such as choosing the differential equation solver. This circuit simulator allows the user to set the parameters of the simulation through the circuit_inputs.csv spreadsheet. A commercial software will usually have a graphical user interface where the schematic file will contain the circuit drawn using a schematic editor. For large circuits, most commercial simulators provide the facility to create subsystems which can contain segments of the circuit. Therefore, the overall circuit will appear as blocks connected together. In this circuit simulator, a circuit is represented on a spreadsheet and saved as a .csv file, and as shown in the beginning of this chapter, the representation of the circuit is in a manner that is as visual as a graphical schematic editor. For a large circuit, if the user wishes to break up the circuit into smaller segments, each subcircuit can be placed in a different spreadsheet and joined to other spreadsheets using jump labels as connectors. The user needs to specify all these spreadsheets as .csv files in the circuit_inputs.csv spreadsheet. Therefore, a similar level of convenience in representing circuits is provided in the circuit simulator.

The circuit simulator lacks a graphical user interface and one may be designed at a later stage. The lack of a graphical user interface may appear to make the simulator inconvenient to use at first. However, for a large circuit, a graphical user interface plays a minimal role as quite often a user tends to use script files. For example, if simulating a circuit with several power converters, it is quite common that the parameters of the circuit are not entered directly into the dialog boxes of the components in the schematic. Rather, a unique variable is entered in these dialog boxes and the value of the variable is assigned in a script file that is executed before running the simulation. In the case of a circuit with embedded elements and subsystems, to continuously open dialog boxes of components and change their value will be inconvenient while changing them in a well-commented script file makes it much easier to keep track of the state of the circuit. For a larger circuit, the greatest challenge is in plotting waveforms of currents, voltages, and control signals. This circuit simulator provides the option of capturing the entire simulation data in separate time intervals such that the user can plot those time intervals that are of importance or plot the entire simulation time duration if that is desired. Many existing simulation software is not targeted toward large circuits, and therefore plotting waveforms is quite often inconvenient.

Chapter 4
Interface for User Control Functions

Abstract This chapter describes how a user can write control functions for a simulation. The chapter describes how the control functions have to be written as Python files and specified in the simulation parameter spreadsheet. Every control function will have an interface to the simulation in terms of inputs and outputs, and this interface is described by a spreadsheet called a descriptor. Besides inputs and outputs, every control function can use certain types of variables that perform special actions. The chapter describes the importance of each type of control variable and how they are implemented in the simulator. The chapter describes how control functions are scheduled by the simulator using time events, and with an example, it is described how the simulator ensures that the control functions execute at the desired time instant. A simple example has been provided to describe how control functions can be interfaced with the simulation and also with each other.

Keywords Control functions · Control interface · Interface descriptors · Static variables · Reserved variable types · Time events · Event scheduling · Nested control

4.1 Introduction

The previous chapter introduced the user interface to the circuit simulator. In particular, the simulation parameter spreadsheet circuit_inputs.csv was described, and it was shown how a row was devoted to specifying the control functions in the simulation. In the previous chapter, the data structure of each type of circuit component was described in the form of tables. The components that could be controlled were described to have ControlTags with control values that can be modified by user-defined control functions. This chapter is completely devoted to user-defined control functions. The word controller and control code will be used for any code that the user wants to implement in the circuit simulation. As already stated before, it is not necessary that this control code be performing a control action and for that matter, there need not be a controllable component in the circuit. This control code could be

S. V. Iyer, *Simulating Nonlinear Circuits with Python Power Electronics*,
https://doi.org/10.1007/978-3-319-73984-7_4

merely a processing tool used by the user in real time as the simulation progresses such as calculating efficiency, harmonic content, root mean square values.

The prime target of the circuit simulator is to simulate circuits with multiple power electronic converters. Therefore, including control functions in a simulation must be convenient to the user as the user may need to design control functions for a number of a converters. This chapter will describe how each control function has a descriptor which defines not only the input–output map of the control function but also defines special variables for the control function. The chapter will describe the need to define special variables and how the user can use them particularly when cascaded or embedded control functions need to be designed. In addition, the chapter will also discuss the concept of time scheduling of control events and also how the circuit simulator can ensure that every control code will run at exactly the intended time instant even when the time step of the controller can be very different from the simulation time step.

This chapter primarily intends to show how the control interface with this circuit simulator is about as effective and easy to use as the control interfaces available with most commercial software. This chapter describes how multiple control functions can be incorporated in a single simulation and how the simulator enables connectivity between control functions. The objective of designing the control interface in this manner is to replicate the implementation of control algorithms on microcontrollers in hardware. In the future, the option might be provided to design control with a particular hardware architecture to decrease the time for implementing the final hardware.

4.2 Inclusion of Control in the Simulator

Chapter 3 described the user interface for designing the circuit and specifying simulation parameters. As was mentioned, the spreadsheet circuit_inputs.csv specified the simulation parameters—circuit schematic spreadsheets, simulation time step, and time duration of simulation. There is also another row for control files. If this row is left blank, the simulator will not process any control files. If the user wishes to specify control files, the user can specify any number of control files in separate cells in that same row. Table 4.1 is an example of how two control files control1.py and control2.py can be specified by the user in circuit_inputs.csv. There is no limit to the number of control files that can be included in a simulation. At the time of writing this book, control can be implemented as Python 2 code and the control files have to listed as .py files. Specifying a control file that does not exist is an error, and the simulator will abort with an error message.

The simulator deals with user control files in a manner that is described by the following block diagram (Fig. 4.1). The user writes code as a .py file. The control code written by the user is shown as files Control1.py, Control2.py to ControlN.py on the left. Each such control file is inserted into a function by the simulator. This is shown on the right. Control1.py is inserted into the function Control1_func, Control2.py is

Table 4.1 Control files in circuit_inputs.csv

Name of control files	control1.py	control2.py

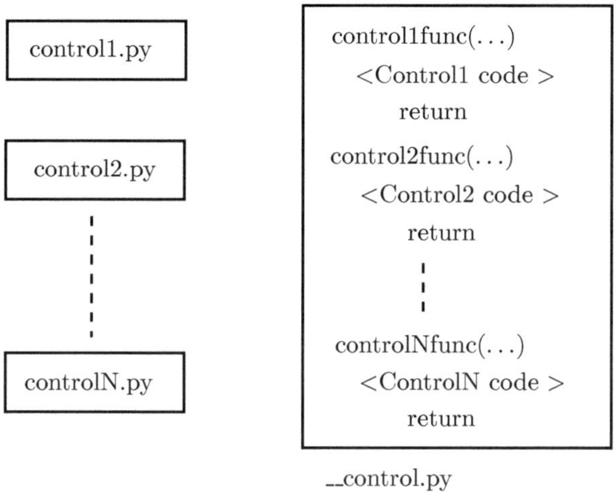

Fig. 4.1 Inclusion of control functions in the simulator

inserted into the function Control2_func, and so on. Essentially, the function which contains a user control file has "_func" appended to the name of the control file. This enables the simulator to execute the control files by evaluating the functions. Since there can be multiple control files in a simulation, the code in each control file will be inserted into a separate function by the simulator. All these functions will in turn be inserted into the file __control.py. Therefore, the user cannot name a control file by this name. The simulator will import the file __control.py, thereby gaining access to all the member functions within it which are the functions containing user control code.

In most commercial software, a controller is usually a block with input and output ports. In a similar manner, it becomes necessary to describe the port connections with this circuit simulator. This is done by a descriptor spreadsheet. When the user wishes to implement a controller, all the user needs to do at the beginning is name the file—for example, let us call a control file by user_control.py. When the circuit simulator is notified of the control file in circuit_inputs.csv, the simulator will look if a descriptor exists for such a file. Any control file will have a corresponding descriptor spreadsheet with _desc.csv appended to the name—for example, for a control file user_control.py, the descriptor file will be user_control_desc.csv. When the user specifies a new control file, a blank descriptor spreadsheet with default input and output will be created by the simulator. The flowchart in Fig. 4.2 describes the process. When the simulator finds a descriptor file, it will use that file to read the latest parameters of each control file. We will now look at the concept of how a control file is handled by the simulator with the following basic example with a single input and single output.

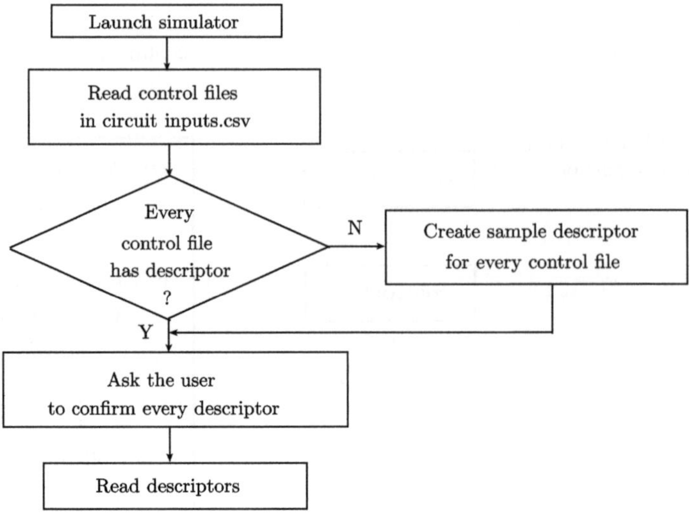

Fig. 4.2 Descriptor for control functions

The inputs to the controller are usually meter outputs—Ammeter and/or Voltmeter. The later section will describe how a controller could be interfaced with another one. The circuit simulator will create a single entry in any new control descriptor file with the structure shown in Table 4.2. These three fields of an input port will be present in a row of any descriptor spreadsheet the simulator creates by default. The first column "Input" is to let the user know that the row corresponds to an input port to the controller. The second column is the meter which serves as the input to the control code. This has to be equal to the component which is the meter. Notice that the entire component as it appears in the spreadsheet should be specified. The third column is for the user to specify how the user wants to access this input in the control code. For the above example, the user can access the measured current output of Ammeter_A1 by the variable curr_input in the control code user_control.py. The simulator will copy the current output of Ammeter_A1 to the variable curr_input and then execute the control code. This variable name is up to the user and can be anything as long as it is a legal variable name and also should be unique within the control code—i.e., it should not be used for another input or output port or any other variable to avoid corruption of data. The circuit simulator will create only one such row above and will extract a meter at random from the circuit. The user can have multiple inputs by adding rows similar to the one above and adding more meters. There is no limit to the number of inputs there are in a control descriptor spreadsheet. The simulator determines a row to be an input port when the first column is "Input".

The next port to be described is the output port. A row in the descriptor spreadsheet corresponding to an Output port will be as shown in Table 4.3. The first two columns are similar to the input port. The word "Output" in the first column tells the simulator that the row corresponds to an output port. The second column is the name of the

Table 4.2 Descriptor entry for control input

Input	Element name in circuit spreadsheet = Ammeter_A1	Desired variable name in control code = curr_input

Table 4.3 Descriptor entry for control output

Output	Element name in circuit spreadsheet = ControlledVoltageSource_Vin	Control tag defined in parameters spreadsheet = Vsource	Desired variable name in control code = Vsource	Initial output value = 0

controllable component. In this example, a ControllableVoltageSource_Vin has been used. The simulator will look for a controllable component in the circuit, and if one is found will insert that into the descriptor spreadsheet as an example for the user to add others. The simulator will insert only one controllable component, and this is chosen randomly. The third column is the control tag of the controllable component. At this point, the reader should refer to Chap. 3 where the ControllableVoltageSource component type is described. Any controllable component type has a parameter called ControlTags. This is a parameter of the component type that can be changed by the user in the parameters descriptor spreadsheet. To begin with this parameter is a list because a controllable component could accept multiple control signals. In the case of the ControllableVoltageSource, the list has only one parameter and that is the desired output voltage of the source. Therefore, ControlTags by default would be ["Voltage"]. This is the default name given by the simulator to the control input of the component of type ControlledVoltageSource in the parameter spreadsheet. As stated in Chap. 3, it is always recommended to change the name of the inputs in ControlTags to something unique and easily identifiable. Let us then, rename the control input to Vsource. This renaming is not done in the control descriptor spreadsheet. It is done in the parameter spreadsheet corresponding to the circuit as the name of the control input is a parameter of the component ControllableVoltageSource_Vin. In the control descriptor spreadsheet, the simulator automatically extracts one of the items in the parameter ControlTags from the component object ControllableVoltageSource_Vin and inserts it into the third column as shown in Table 4.3. It should be noted that in case a component does have multiple control inputs and ControlTags is a list with multiple entries, the entries would have to be made in separate rows of the descriptor spreadsheet. A single row of a descriptor spreadsheet will only connect a single control input of a controllable component to a variable inside the control code. The simulator will only extract one control tag of a controllable component to serve as an example to the user. The fourth column will ask for the name of the variable that the user wishes to refer to the control input by. In the above example, it has been used as Vsource. Therefore, any value assigned to the variable Vsource in the control code will automatically be transferred to the Vsource ControlTag in

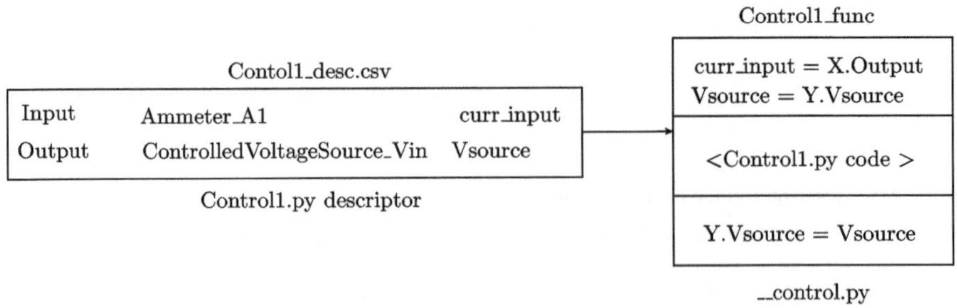

Fig. 4.3 Input and output ports in a control function

ControllableVoltageSource_Vin. The last column is an initial value. This will be the value of the output when the simulation starts.

The block diagram of Fig. 4.3 describes how the simulator uses the control descriptor spreadsheet. Let us consider a user file Control1.py. This control file will have the descriptor spreadsheet Control1_desc.csv. If it exists, the simulator will read the parameters of the control file, and if it does not exist, the simulator will create a blank spreadsheet for the user to change. On the right is the function Control1_func into which the control code within Control1.py is inserted by the simulator. Let us consider the basic example above of a single input port and a single output port. On the right, it is shown how the input and output ports are made available to the user. The example above considers an Input port which is fed by an Ammeter_A1, and the variable by which the user can access the Ammeter measured current is through the variable curr_input. The simulator assigns to the variable curr_input the current output of the Ammeter_A1 inside the function Control1_func before the user code with the statement:

```
curr_input = X.Output
```

Here, X is the component object corresponding to the Ammeter_A1. This concept has been explained in Chap. 3. Each component in the circuit is an object created by instantiating a class of that component type. Therefore, in the above statement, X is the object created by instantiating a class of the Ammeter type for the component Ammeter_A1. Chapter 3 has the details of how this is done. What needs to be emphasized here is that the simulator automatically extracts the measured output of Ammeter_A1 by accessing the object corresponding to Ammeter_A1 and assigning it to curr_input. Therefore, the subsequent user control code which uses curr_input now contains the measured current of Ammeter_A1.

In a similar manner, the output is also interfaced. The only difference being that Output ports are interfaced twice—before the user control code and after the user control code. The variable Vsource is made available to the user to access the ControlTag Vsource of ControllableVoltageSource_Vin. The user control code will therefore change the variable Vsource within control code using any set of statements. This variable Vsource will directly alter the ControlTag of the controllable component specified in the descriptor spreadsheet—in this case, the ControlTag is Vsource. This is done using the last statement:

```
Y.Vsource = Vsource
```

In this case, Y is the object produced by instantiating the class ControllableVoltage-Source for the component ControllableVoltageSource_Vin in the circuit. Therefore, any change in the user variable Vsource will automatically be transferred to the controllable component. The reverse statement is included before the control code in the function:

```
Vsource  = Y.Vsource
```

This is to provide the user with the updated value of the ControlTag of the component in case another control function also changes the value of the ControlTag.

4.3 Special Variables in Control Code

As described in the previous section, any control code written by the user and specified in simulation parameters spreadsheet circuit_inputs.csv will be inserted into a function. So control code in Control1.py will be inserted into the function Control1_func. All the control functions will be written in the file __control.py. By doing so, the file __control.py can be imported by the simulator and each control function can be executed using the "eval" function.

The previous section described how a basic controller can be designed with one input and one output port. However, with just input and output ports and no other special functionalities, only extremely simple controllers can be designed. As an example, the user can use a number of variables for various mathematical operations. In Python, it is not necessary to declare a variable. A variable comes into existence the first time it is used. However, in the context of a function, a variable exists only within the function. The variable is created when the function is called and is destroyed when the function is terminated. Any variable used by the control code will not store its value between iterations and will always start with default values. In some cases, this may be a problem. Consider the case of an integrator. The user wishes to integrate the measured current curr_input in the previous example. The expression for that would be:

```
curr_integ = curr_integ + curr_input*dt
```

As is fairly obvious from the above equation, the integrator is based on storage. It adds the new value curr_input*dt to the stored value of the integrator - curr_integ. If curr_integ is initialized to a default value (say zero) in the beginning of the control code, when the function is executed at every iteration, it will always start at this default value and add only the latest integrator input. This is not the desired operation of the integrator. The integrator value should be initialized to a default value once in the beginning of the simulation and after that should accumulate the latest integrator

Table 4.4 Descriptor entry for StaticVariables

StaticVariable	Desired variable name in control code = curr_integ	Initial value of variable = 0.0

inputs. In order to do this, we need to store the integrator output between iterations of the simulation. This can be done with StaticVariables.

A StaticVariable is a parameter of a controller. This implies it has to be defined in the control descriptor spreadsheet. Table 4.4 is a typical entry for a StaticVariable in a control descriptor spreadsheet. As an example, let us consider the variable curr_integ above. The first column StaticVariable tells the simulator that the row describes the parameters of a StaticVariable. The second column is the name of the variable the user wishes to use for this static variable. In a manner similar to the Input and Output ports, the user can use the defined StaticVariable in control code and the simulator will ensure that the latest value of the StaticVariable is copied into it before the user code begins. The third column is the initial value of the StaticVariable. The default is zero, but it can be any finite number. Similar to ports, a control code can have any number of StaticVariables. StaticVariables need to be unique within a control code, but they can be repeated in other control codes. The simulator will keep StaticVariables in a control code separate from the StaticVariables in other control code. Therefore, StaticVariables are local to a control code and cannot be used to share data between controllers. It is recommended that a user declare the variables in user control code as StaticVariables unless a variable is a constant or some other basic parameter which is initialized at the beginning of the control code. In that case, the variable will be assigned a value every time the user control function is evaluated.

StaticVariables within a user control function are dictionary items. In the above example of the integrator, the StaticVariable curr_integ is in a dictionary with another StaticVariable pi_output:

```
{``curr_integ'': 0, ``pi_output'': 0}
```

The keys of the dictionary are the names of the StaticVariable. This ensures that a StaticVariable within a file is unique as two keys in a dictionary cannot be identical. StaticVariables need to be unique within a control function to prevent data corruption. However, identical control code can be present in different user control functions, and this implies identical StaticVariables in different user control functions. This functionality could be needed when a circuit could have modular but identical blocks where each block is controlled by identical user control functions where the only difference may be the inputs and the outputs. To achieve this, each user control function has its own StaticVariable dictionary. The dictionary structure makes it convenient to insert the variables into the user control functions and also to extract the updated values at the end of the control function as shown in the block diagram of Fig. 4.4. Since the keys of the StaticVariable dictionary are the desired names of the StaticVariables to be used in the control code, the assignments before and after the user control code ensure that the latest values of the StaticVariables are

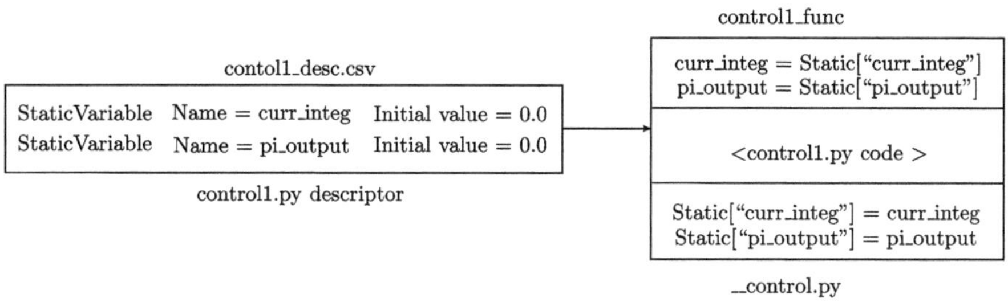

Fig. 4.4 StaticVariables in a control function

passed back and forth between the user control functions. These assignments are done automatically by the simulator in a manner similar to the previous cases.

As stated before, StaticVariables are defined for every user control function and the StaticVariables defined for one are not accessible in another. In the previous section, Input and Output ports for a control function were described. However, the input ports are typically the measured values of meters like Ammeters and Voltmeters while the Output ports are control signals to ControlTags of controllable component objects. However, at times, control files need to be connected together—the output of one control file is the input to another. Complex control schemes may have several stages of cascaded control or embedded control. Another aspect that is not dealt with using StaticVariables or Output ports is that in order to debug a controller, it may be necessary to plot a control variable. In order to do so, the control variable needs to be written to the output data file and therefore access outside the user control function needs to be provided. Another type of variable is provided for both these tasks and these are called VariableStorage. VariableStorage objects have two properties - they are made available to all control functions that the user defines and they have a field specifying whether they should be plotted in the output data file. The parameters of a VariableStorage type are listed in Table 4.5. The VariableStorage in the first column shows the circuit simulator that the variable is of the type of VariableStorage. The second column is the name of the variable that the user wishes to use in the control code. Since, this is a stored variable, an initial value is needed to ensure that a random garbage value is not generated. The last column is the column that asks the user whether the variable should be written in the output data file so that it can be plotted by the user. If the user says "Yes" as in the above case, the variable will be written to the output data file while if the user says "no", the variable will not be written in the output data file. In either case, a variable of type VariableStorage, will be made available across all control functions, irrespective of which control file descriptor contains the definition. On the contrary, once a variable has been defined in a control file descriptor spreadsheet, a repeat definition in another control file descriptor spreadsheet is an error and the simulator will abort with an error message.

The method of implementation of VariableStorage is different from other variable types. Instead of having different dictionaries for each user control function, there is a single dictionary for all the user control files in a simulation. By defining a

Table 4.5 Descriptor entry for VariableStorage elements

VariableStorage	Desired variable name in control code = plot_variable1	Initial value of variable = 0.0	Plot variable in output file = yes

single dictionary and making it available to all user control functions, the sharing of data contained by the VariableStorage elements is possible across all user control functions.

```
{''plot_variable1'': [0.0, ''yes'']}
```

The above dictionary shows how the VariableStorage element in the example above is defined. The key is equal to the name of the VariableStorage object while the value is a list with two elements—the first being the value of the VariableStorage element and the second being the flag whether the element should be written to the output data file. Figure 4.5 will show how VariableStorage elements are used by the simulator. The VariableStorage element "plot_variable1" above could have been defined in the control descriptor spreadsheets of any one of the control files. However, as shown, it is made available in each user control function before the user control code and also extracted from each user control function at the end of the control code. As can be seen from the block diagram of Fig. 4.5, VariableStorage is as extremely powerful variable type which is similar to a global variable in other programming languages. It could provide a great deal of flexibility in control design by allowing exchange of control variables. However, there is a great potential of data corruption as variables can be changed in every control function. Therefore, the use of this VariableStorage element should be limited to writing control variables to the output data file and for connection between control files. The use of VariableStorage elements as replacements for regular StaticVariables is not recommended as the risk of data corruption is high. The example provided in Chap. 5 will describe the recommended use.

With respect to these two variable types described above, there is a special property related to the Python programming language that the user can use. When a variable is declared as StaticVariable or VariableStorage, by default it is a floating point number whose default initial value is 0.0. The user can change this default initial value to any other floating point number. However, in Python, every variable is an object of a particular type. The object corresponding to a variable can be changed by initializing it to that type. To elaborate, let us examine a variable "x":

```
x = 0.0
```

In Python, a variable does not have to be declared. Therefore, the above statement will create an object of floating point type and assign that to x. However, if another statement is written:

```
x = [0.0, 0.0, 0.0]
```

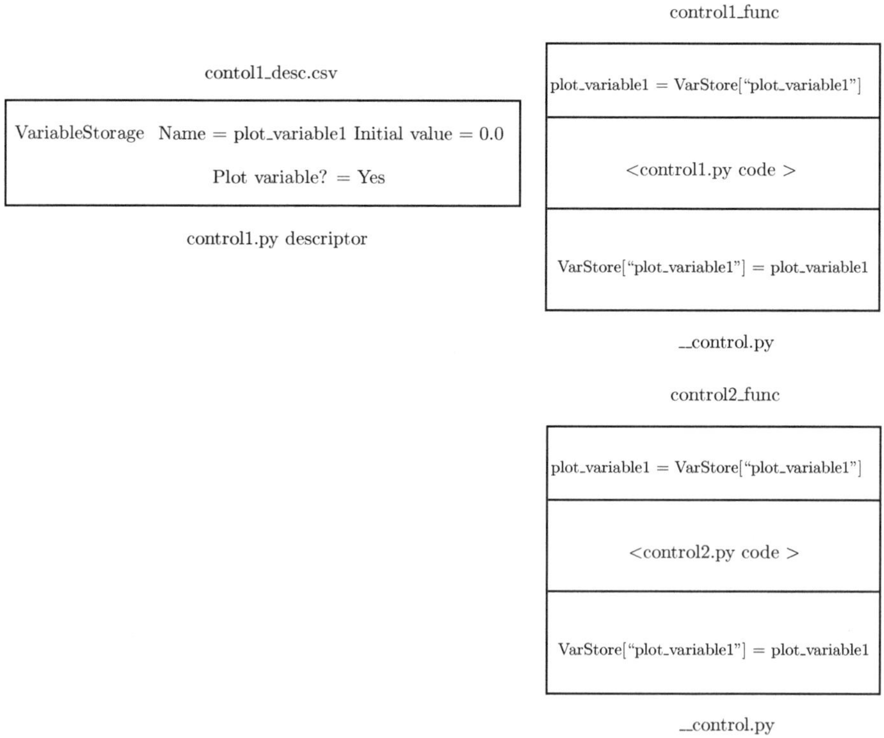

Fig. 4.5 Variable storage types in control functions

This will destroy the object of floating point type and create another object of type list and assign that to x. This facility in Python can be used to create other objects for StaticVariables and VariableStorage. For example, the StaticVariable "curr_integ" defined above and initialized to 0.0 in the control descriptor can be changed by the user within the control code to:

```
curr_integ = [0.0, 0.0, 0.0]
```

By writing the above statement in the control code, the StaticVariable dictionary for the control function will have the key "curr_integ" with the above list as its value rather than the floating point number. This is a particularly useful concept if control code needs to be written in matrix form for polyphase systems. The above variable curr_integ could now contain the integrated values of the currents in phase a, phase b, and phase c of a three-phase power system.

Table 4.6 Descriptor entry for TimeEvent variables

TimeEvent	Desired variable name in control code = t1	First time event = 0.0

4.4 Time Scheduling Control Code

The previous sections described the input/output structure of control in the simulator and also the special types of variables that can be used—StaticVariables and VariableStorage. In this section, time scheduling of control functions will be described. A control function is very rarely required to execute at the same time step as the simulation. In most cases, the time step of the control function is much larger than the simulation time step as this is how the control will be implemented in a microcontroller or microprocessor, and for real-time control, practical values of control frequency need to be chosen. In some cases, the control function needs to be run at a fixed time step due to the nature of control—for example, in a resonant converter, the control has to be timed with respect to the resonance of the circuit.

Every control function defined by the user will be provided with the current time instant of simulation through the construct t_clock. This variable can be accessed in every control algorithm by the user with the simulator updating this variable with the latest time instant of simulation before the user control code. In order to schedule the control code, the simulator has a special construct called a TimeEvent. The parameters of a TimeEvent are listed in Table 4.6. As before, the TimeEvent in the first column tells the simulator these are the parameters of a time event variable. The second column is the desired name in the control code. The user can use this variable to assign control events at future time instants, and this variable will be stored and used by the simulator to ensure that the control code executes at that time instant. The third column is the first time event that needs to be scheduled. The default is the start time 0.0 but can be changed to anything. This time event is typically used in conjunction with t_clock in the manner described below:

```
if t_clock>t1:
        ____________
        Control code
        ____________
        t1 = t1 + t1_period
```

By inserting the control code within a conditional statement that checks if t_clock is greater than the time event t1, it is possible to adjust t1 using t1_period to ensure that the control code will run only every t1_period. Moreover, t1_period need not be a constant and can be a variable based on the control code.

A control function can have any number of TimeEvent variables. Every control function has its own dictionary that stores the TimeEvent variables defined in the descriptor as follows:

```
{''t1'': 0.0, ''t2'': 0.0}
```

It is completely up to the user to generate values for the TimeEvent variables in any way desired. Since there is a separate dictionary containing TimeEvent, it is possible to define identical TimeEvent variables for multiple control functions. However, it is very important for the user to update every TimeEvent in the control code. Failure to do so will result in the TimeEvent not changing. As will be described soon, this will cause the control code to execute at the rate of the fastest among the remaining control functions or the simulation time step, whichever is faster. Usually, neither of these options are what the user wishes for the time step of executing control functions. Therefore, it is recommended that the user makes sure that any TimeEvent defined for a control function is updated correctly. The advantage of defining TimeEvents is that control functions can be executed at any arbitrary time instant in the simulation. As said before, not only can the time period of control functions be greater than the simulation time step but they can also be smaller. Before explaining this concept, the link between the control functions and the simulator will be described.

The question is how are the simulator circuit analysis blocks—loop analysis and nodal analysis—linked to the control functions? In the absence of control functions, the circuit analysis blocks would run at the simulation time step. When the simulation contains control functions, the circuit analysis blocks will run either at the simulation time step or when a control function generates an "event". A control function generates an event when any of its output ports changes. There is no threshold for change; if the value of the output port in the current execution of the control function differs from the value before the execution, an event is generated. When an event is generated by one of more control functions, the circuit analysis blocks are executed. To elaborate on how this is coordinated, let us consider the following example in Fig. 4.6. In this example, the simulation has three control files. In this section, the details of the control files are not important. The emphasis is on the time scheduling of these control functions. As shown, the first control function is called reference voltage loop. This is the outermost control loop and is a slow control loop that has a time period of 1 ms. The second loop is the voltage control loop which is a faster inner control loop with a time period of 100 μs. The third and the innermost control loop is the pulse width modulator which has a time period of 100 ns. The simulation time step has been chosen to be 5 μs. This is a typical control layout for many power electronics simulations—1–5 μs is a reasonably small time step for a power

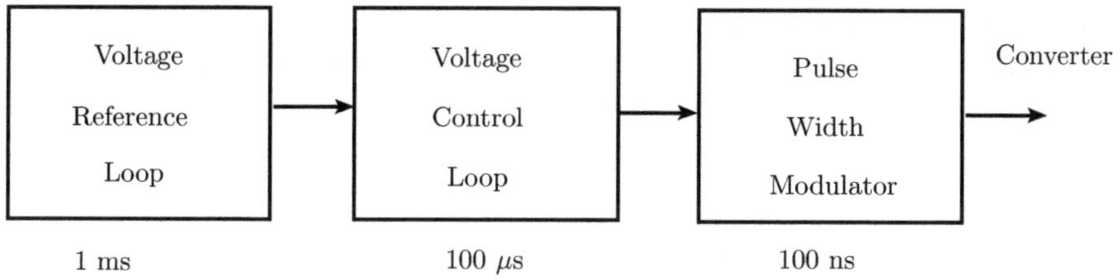

Fig. 4.6 Nested control to describe timing

converter switched between 1 to 10 kHz. The voltage control can be at a time period much higher than the simulation time step. However, the Pulse Width Modulator is a hardware implementation detail. This is achieved using specialized hardware on microcontrollers or can be achieved using analog comparators. In either case, the resolution of the Pulse Width Modulator is very high and therefore the time period of this control block may have to reflect a hardware detail. For a 10 kHz converter, the switching time period is 100 µs, for which the Pulse Width Modulator time period of 100 ns results in 1000 samples in a switching period. In most cases, this results in sufficient accuracy for generating switching signals. The time period can be reduced further if the converter is a resonant converter or if the switching frequency can be higher.

The significance of the above example is that there are now four different time periods in the simulation—three from control functions and the fourth from the simulation time step. The time periods mentioned are just examples and can change. However, it is important to note that there are control time periods that are greater than the simulation time step and control time periods that are smaller than the simulation time step. To begin with, each of the control functions can be called Volref.py, Volcon.py, and Pwm.py. Therefore, there are descriptors for each control function - Volref_desc.csv, Volcon_desc.csv, and Pwm_desc.csv. The TimeEvent parameters in each control descriptor spreadsheet are provided separately in Tables 4.7, 4.8, and 4.9. The TimeEvent variables could have been declared to be the same as TimeEvents are local to a control function and are not global like the VariableStorage variables. However, for clarity in examining the time scheduling of these control functions, distinct variables are chosen. Each control function will have a separate TimeEvents dictionary:

```
{``tvolref'': 0.0}
{``tvolcon'': 0.0}
{``tpwm'': 0.0}
```

Let us assume that the control files will update their TimeEvent variables according to the time periods discussed before. Therefore, the following statements will be present in the respective control files,

```
Volref.py : tvolref = tvolref + 0.001
Volcon.py : tvolcon = tvolcon + 100.0e-6
Pwm.py :    tpwm = tpwm + 100.0e-9
```

Table 4.7 TimeEvent in Volref_desc.csv

TimeEvent	Desired variable name in control code = tvolref	First time event = 0.0

Table 4.8 TimeEvent in Volcon_desc.csv

TimeEvent	Desired variable name in control code = tvolcon	First time event = 0.0

Table 4.9 TimeEvent in pwm_desc.csv

TimeEvent	Desired variable name in control code = tpwm	First time event = 0.0

Refer to the conditional statement shown above where the time event was compared with the time instant of simulation t_clock, to know how the TimeEvent update takes place and control code can be executed once within the time period.

With the above background, let us examine the process of scheduling. The simulation creates a list called TimeVector. Let us start from a random time instant when the simulation execution runs completely. This implies circuit analysis—loop and nodal analysis as well as execution of all control functions. At this point of time how does the simulation proceed? When a control function is evaluated, it may so happen that the control code is not executed due to time instant of simulation t_clock being less than the TimeEvent for that control function. For example, let us say that the current time instant of simulation t_clock is 80 μs while Volref.py has a TimeEvent of 1 ms. Since t_clock<tvolref, the control code in Volref.py will not execute and execution will only take place when t_clock is 1 ms. Evaluation of all control functions will result in values of their TimeEvents remaining unchanged or updated to new values according to their time periods. These values of TimeEvents are added to the TimeVector list. Therefore, after evaluating all the control functions, TimeVector will be:

```
TimeVector = [tvolref , tvolcon , tpwm]
```

The list contains the values of the TimeEvents. To this list is now added the simulation time instant which we called tode—meaning time instant of the Ordinary Differential Equation solver. So, TimeVector is:

```
TimeVector = [tvolref , tvolcon , tpwm, tode]
```

The above list is arranged in ascending order such that the smallest time value is the first element. This smallest value will now be the next instant of evaluation.

The word “evaluation” has been used instead of “execution”. This is because the simulation now checks for another status. Did any of the control functions generate an event? As described before, a control function is said to generate an event when one or more of its outputs changes. There are no minimum or maximum requirements for a change in the output. Any change in the output however small or large will trigger an event. If one or more control functions generate an event, the simulator will now go through the entire simulation cycle—loop and nodal analysis as well as control functions. This is because when a control function generates an event, this means an output has changed which in turn causes a change in the circuit and therefore the circuit needs to be solved again completely. To compare this with the other possibility, what happens when no control function generates an event because no control function experiences any change in any of its outputs? In this case, the simulator will only evaluate the control functions at the next time instant which is

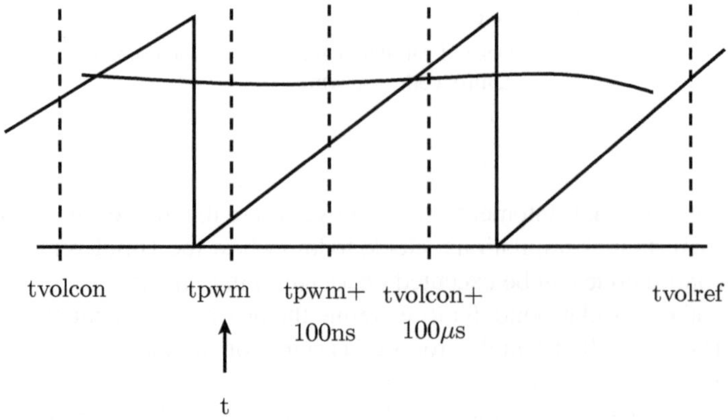

Fig. 4.7 Timing diagram of control functions

the smallest time instant in TimeVector. The following example will describe the difference between simulation evaluation and simulation execution. Let us consider the time instant of simulation marked as t in Fig. 4.7. As is evident, this time instant was chosen because tpwm was the smallest in TimeVector. TimeVector can be:

```
TimeVector = [tpwm, tvolcon+100.0e-6, tvolref]
```

Let us neglect the time instant from the simulation time step to focus on the effect of control functions. Let us suppose that at the instant t shown, none of the control functions generated an event. Only Pwm.py updated its TimeEvent from tpwm to tpwm+100.0e-9. The remaining TimeEvents in the other two control functions are the same and are shown in the figure. So TimeVector is:

```
TimeVector = [tpwm+100.0e-9, tvolcon+100.0e-6, tvolref]
```

Since none of the control functions has generated an event, there is no point in running the circuit analysis. There has been no change in the circuit, and the time difference between the tpwm and tpwm+100.0e-9 is 100 ns. This is much less than 5 μs which is the simulation time step. The simulation time step is chosen with respect to stability of the simulation. As will be described in Chap. 7, if the simulation time step is not small enough as compared to the time constant of the branches in the circuit, the simulation can become unstable. However, when the simulation updates the time instant from tpwm to tpwm+100.0e-9, the simulation is checking for control events that change the state of the circuit. Simulating at a potential time step of 100 ns when 5 μs has been found to be sufficient will increase the computational burden of the simulator significantly and slow it down. Therefore, if no event has been generated at t=tpwm, at the next time instant t=tpwm+100.0e-9, it is only necessary to evaluate the control functions. The simulator will evaluate all control functions; however, by using the conditional time check shown above, only Pwm.py will be evaluated. Any control function may generate an event in any time instant of simulation or it may so happen that no control function may generate an event at all for a number of these time instant updates.

Let us examine two possibilities. The first possibility, suppose the control file Pwm.py generates an event. This could be a change in one of its outputs which may be the switching signal to the converter—turning on or off a Switch. When an event occurs, the state of the circuit has changed. In this case, the resistance of a branch has changed and in this particular case, the resistance of a branch can change by a huge order—a few milliohms to several megaohms. For accurate simulation that resembles hardware, it is essential that the simulator now executes all the circuit analysis functions and updates the currents and voltages and all other variables in component objects in the circuit. As before, all control functions are also executed and TimeVector is updated. The next time instant of simulation will be the smallest element of TimeVector.

The second possibility is that the smallest time instant may be generated by Volcon.py, i.e., tvolcon shown in the figure above. As an example, let us assume in this case that Volcon.py does not produce an Output but only generates the duty cycle or modulation signal for the pulse width modulator. Therefore, Volcon.py does not affect a circuit component directly in the way Pwm.py does though it affects another control function which is Pwm.py. The duty cycle generated by Volcon.py is of the type VariableStorage as the duty cycle has to be accessed in Pwm.py. Details of the control will be provided in the next section. A change in one or more variables of the type VariableStorage will not generate an event in the simulator. This is because VariableStorage is used for interfacing between control functions and for writing control variables to the output data file. VariableStorage types will not directly impact the state of the circuit the way Output types do. Therefore, if tvolcon is the smallest in TimeVector, it is ensured that the simulator will evaluate the control function at that particular time instant. However, circuit analysis will not be performed without an event being generated.

In the above description, the time step of the simulator was left out to focus on the effect of TimeEvents generated by the control functions. However, even in the presence of control functions, the time step of the simulator can decide the next time instant of simulation. As stated before, the time instant of simulation is always added to TimeVector.

```
TimeVector = [tpwm, tvolcon+100.0e-6, tvolref , tode]
```

The smallest time instant in TimeVector will be the next simulation time instant. In case tode is the smallest, the simulator will execute the entire cycle even if no event is generated by a control function. This is for two reasons. In case there are no control functions, TimeVector will only contain tode and therefore, the time step of simulation alone will decide the next instant of circuit analysis. Therefore, the simulator must execute all circuit analysis functions as this will be a case of a fixed time step circuit simulation which is quite often the case when simulating passive circuits. The second reason is that when tode is the smallest time instant in TimeVector even when control functions are present, it may be possible that the control functions have significantly larger time periods than the simulation time step. In this case, the simulator must execute all circuit analysis functions when tode is found to be the smallest as failure to do will result in the simulation being unstable. As stated below, a simulation at every simulation time step is a necessity for stable simulation—that is how the simulation time step is chosen. However, when tode is found to be the smallest time instant in TimeVector, it is necessary to update tode with the simulation time step. So,

```
tode = tode + dtode
```

where dtode is the simulation time step specified by the user in circuit_inputs.csv.

4.5 Interfacing Control Code

The previous section described how the control functions are scheduled and how the circuit simulator interfaces the control functions with the circuit analysis functions. However, the control functions shown in the previous example were treated as black boxes with an emphasis only on how the TimeEvents are generated. In this section, it will be described how the control functions are interfaced. Detailed control algorithms will not be presented. The focus will be on showing cascaded control can be designed with the circuit simulator.

Let us start with the design from the inner most control function to the outermost. Therefore, first Pwm.py. The parameters in Pwm_desc.csv will be as listed in Table 4.10. Let us assume there is only one Switch called S1 in the converter - for example, a simple buck converter. This Switch_S1 has a control tag named as S1_gate in the circuit parameters spreadsheet. This control tag is accessed by S1_gate in Pwm.py. Since a pulse width modulator is being programmed, a carrier wave will need to be generated. A StaticVariable called carr_signal is defined for this purpose. The duty cycle will be the output of the controller which regulates the voltage of the buck converter, and this code is found in Volcon.py. The duty cycle is defined as a VariableStorage type as it is an input from another control function. This could have been defined in the descriptor spreadsheet of Volcon.py, but let us define all the variables of Pwm.py first. Lastly, the TimeEvent tpwm as already described in the previous section.

The code in Pwm.py will be similar to this without going into the details:

```
if t_clock >tpwm:
    carr_signal += (1/5000)*100.0e-9
    if (carr_signal >1):
        carr_signal = 0
    if (duty_cycle>carr_signal):
        S1_gate = 1
    else:
        S1_gate = 0
    tpwm += 100.0e-9
```

Table 4.10 Pwm_desc.csv

Output	Element name in circuit spreadsheet = Switch_S1	Control tag defined in parameters spreadsheet = S1_gate	Desired variable name in control code = S1_gate	Initial output value = 0
StaticVariable	Desired variable name in control code = carr_signal	Initial value of variable = 0.0		
TimeEvent	Desired variable name in control code = tpwm	First time event = 0.0		
Variable-Storage	Desired variable name in control code = duty_cycle	Initial value of variable = 0.0	Plot variable in output file = yes	

A brief description of the code is as follows. The block of code executes only when the time instant of simulation t_clock is greater than TimeEvent tpwm. The carrier waveform is a saw tooth waveform of unity magnitude and of frequency 5 kHz as shown in the previous section. Therefore, (1/5000) is the slope of the waveform. This needs to be multiplied by the time period of the modulator, i.e., 100 ns and finally needs to be limited to unity. The next step is the comparison between the duty_cycle and the carr_signal, and this generates the output S1_gate which is connected by the simulator to the control tag S1_gate of Switch_S1. Finally, the TimeEvent tpwm is updated by 100 ns. A few things to note. The input is duty_cycle. Since this variable is of type VariableStorage, it can be accessed in any control function. The next control function will describe how this duty_cycle is produced. The carr_signal variable is a StaticVariable, and therefore, it can be directly accessed within Pwm.py and is updated by a += since the stored value is made available in the function by the simulator. Similarly, the updated value is extracted by the simulator and stored in a dictionary for the next iteration.

Now, the next level of control—Volcon.py. This function contains the controller which we have chosen to be a proportional–integral (PI) controller. This control generates the duty_cycle variable used in Pwm.py. The parameters in Volcon_desc.csv are listed in Table 4.11.

```
if t_clock>tvolcon:
    volt_error = volt_ref - volt_output
    volt_integral += volt_error*100.0e-6
    duty_cycle = 0.001*volt_error + 0.01*volt_integral
    if duty_cycle > 0.98:
        duty_cycle = 0.98
    tvolcon += 100.0e-6
```

Table 4.11 Volcon_desc.csv

Input	Element name in circuit spreadsheet = Voltmeter_Voutput	Desired variable name in control code = volt_output	
StaticVariable	Desired variable name in control code = volt_error	Initial value of variable = 0.0	
StaticVariable	Desired variable name in control code = volt_integral	Initial value of variable = 0.0	
TimeEvent	Desired variable name in control code = tvolcon	First time event = 0.0	
Variable-Storage	Desired variable name in control code = volt_ref	Initial value of variable = 0.0	Plot variable in output file = yes

The purpose of the control is to regulate the output voltage to the reference voltage volt_ref. This reference voltage is generated by the outer control Volref.py and described next. 0.001 is the proportional gain while 0.01 is the integral gain. These values are purely arbitrary and just examples to show how to write control code. This control function contains the PI controller which regulates the output voltage and therefore takes the measured voltage as an Input. It is assumed that the circuit schematic contains a Voltmeter called Voltmeter_Voutput. The measured voltage is made available in the control function by the simulator as the variable volt_output. The error in the voltage and the integral of the error are defined as StaticVariables. The variable volt_error does not have to be defined as StaticVariable as it is calculated within the control code and used immediately. However, it is a safe practice to define as many variables as StaticVariables rather than using the default Python objects to ensure control can be tracked. To elaborate on this, if volt_error was not a StaticVariable, it would not exist if the code block was not executed. Therefore, if for control debugging, the variable is accessed outside this block, the simulator would exit with an error as it is accessing a variable which does not exist. However, when a variable is defined as a StaticVariable, it always has a value even if that value is the initial value specified in the descriptor spreadsheet. The duty_cycle is generated according to PI control. As mentioned before, the duty_cycle is made available in all control functions once it has been defined in one of them—in this case Pwm_desc.csv. Also, duty_cycle does not have to be defined as it is already defined in Pwm_desc.csv, and to redefine it is a violation for which the simulator will abort with an error.

Finally, the outermost control function Volref.py. This function will generate the voltage reference for the previous control function Volcon.py. Volcon_desc.py contained the definition of volt_ref as VariableStorage. The parameters in Volref_desc.csv are listed in Table 4.12.

Table 4.12 Volref_desc.csv

StaticVariable	Desired variable name in control code = volt_setpoint	Initial value of variable = 100.0
TimeEvent	Desired variable name in control code = tvolref	First time event = 0.0

```
if t_clock>tvolref:
    volt_ref += 0.01
    if volt_ref>volt_setpoint:
        volt_ref = volt_setpoint
    tvolref += 0.001
```

This control function generates the reference voltage to be used by Volcon.py as a gradual ramp which is clamped at volt_setpoint. volt_setpoint is defined a StaticVariable with the initial value as 100. Therefore, Volref.py ensures that the buck converter starts up gradually with a steadily increasing reference voltage.

This section has described how three control functions in a simulation can be interfaced with VariableStorage and how these variables can be accessed in each control function. It should be noted that VariableStorage can be defined in any control function descriptor. This example was fairly simple, and the purpose was to provide an introduction to writing control functions. A far more elaborate example will be provided in the next chapter.

4.6 Conclusions

Since this circuit simulator is targeted toward power electronics applications particularly with multiple converters, user-defined control functions are an extremely critical component. It is extremely important that a user be able to integrate control into a simulation with the same ease as in a commercial simulator. This circuit simulator allows a user to define multiple control functions with no limits on the number of files. By providing the facility of StaticVariables, the user can implement higher order control functions with complex mathematical calculations. In order to develop control in a modular manner, the VariableStorage type has been provided for the user to be able to break up their control algorithms into modules and interface those modules. The only visible disadvantage is that all the control functions have to be developed with code unlike a commercial software where control functions can be developed by connecting blocks from an in-built library. However, for most complex circuits that are intended for hardware implementation, control functions are developed as code to be programmed in microcontrollers or other forms of embedded controllers.

In a complex circuit with multiple converters, control algorithms can be fairly complex. One of the major challenges with control functions is the time of execution. As an example, in hardware, a control function may be an interrupt service routine connected to a timer. The timer may be configured to generate an interrupt at a constant time period, or the time period may be calculated and loaded into the timer after every iteration. The interrupt service routine ensures that the control function will be executed at the desired time instant. For the simulation to match the hardware implementation, it is essential that the simulator provides a guarantee that every control function will execute at the exact time instant that is desired. Moreover, a user should not have to adjust the simulation time step in order to ensure that it matches the control function time step. In this circuit simulator, a time scheduler ensures that all control functions are evaluated at the desired time instant by providing the variable TimeEvent. The user can specify a TimeEvent for every control function and update this TimeEvent with any time period that is not necessarily multiples of each other or even multiples of the simulation time step. In this manner, the simulator provides accurate time resolution comparable to hardware implementation in a form that is convenient to the user.

This chapter and the previous chapter have described two aspects of the user interface. The first being how the circuit is represented and how the parameters of the circuit components are updated. The second being how control functions are implemented. Chapter 5 will describe in detail how a circuit can be simulated. Moreover, Chap. 5 will provide the user with an example on how a circuit can be developed in stages and how control can be developed and tested in a modular manner. The circuit chosen in Chap. 5 is such that all the functions available with the simulator are used in simulating it. Chapter 5 will show how the interface provided in this circuit simulator is sufficient to design fairly complicated power electronic circuits. Moreover, writing control functions with the user interface described in this chapter provides a powerful platform which is also simple to use at the same time. The reader is encouraged to Switch between Chaps. 3, 4, and 5 while reading Chap. 5.

Chapter 5
Case Study—Shunt VAR Compensator

Abstract This chapter describes how a user can simulate a shunt-connected three-phase VAR compensator realized using a two-level voltage source converter in a three-phase system. The voltage source converter consists of controllable ideal Switches that are turned on and turned off by pulse width modulation. The chapter describes how the user can build this simulation in stages such that every new subsystem added to the circuit can be verified. The chapter also describes how the user can write control functions with detailed examples of each control function in the simulation and also design the control interfaces through descriptors. Every stage of the chapter contains simulation results to show how the project develops. Through this example, every feature of the simulator has been described with details so that users can develop their own simulations.

Keywords VAR compensation · Power quality · Voltage source converter · Pulse width modulation · Proportional–integral controllers · Phase-locked loops · Time scheduling

5.1 Introduction

Chapter 3 described the user interface for generating the circuit schematic and modifying parameters of the circuit components. Chapter 3 described how this user interface modifies the data structures of objects corresponding to circuit components. Chapter 4 describes how a user can implement control functions in a simulation. Chapter 4 describes the interface for specifying the parameters of every control function and how the user can interface control functions, develop complex

---

Electronic supplementary material The online version of this chapter (https://doi.org/10.1007/978-3-319-73984-7_5) contains supplementary material, which is available to authorized users.

S. V. Iyer, *Simulating Nonlinear Circuits with Python Power Electronics*,
https://doi.org/10.1007/978-3-319-73984-7_5

mathematical control, and generate time events. This chapter will use these two chapters and describe how a complex circuit with modular control can be simulated. This chapter therefore serves as a tutorial for a user who wishes to simulate power electronic circuits with this simulator. The description will cover every aspect of the simulation—drawing the schematics, modifying parameters, writing control functions, and developing the simulation in stages.

The main objective of this chapter is to demonstrate how this circuit simulator allows a user to model circuits with power electronic converters with the same ease and flexibility as most commercial simulators. The example will show how a circuit can be built up modularly by adding and removing circuit schematics in the simulation parameter spreadsheet circuit_inputs.csv. The new schematics need to be connected to the existing schematics with jump labels. This is similar to how in a commercial simulator, subsystems can be added to a circuit and the ports of the subsystem connected to the rest of the circuit. Similarly, the control is designed step-by-step and at each stage, the control output is checked by plotting control signals. Therefore, it will be shown how a fairly complex control algorithm can be simplified by breaking it into modules and validating them separately.

The example chosen is a three-phase voltage source converter (VSC) connected to a three-phase grid. To begin with only the three-phase system without the compensator is simulated and a grid synchronization control function is implemented [2]. In the next stage, the references for the shunt compensator are generated with another control function that uses the grid phase angle as input [3]. With these two in place, the compensator is tested as mere controllable voltage sources to test a closed-loop current control scheme [4]. Finally, the shunt compensator is implemented by the VSC and a pulse width modulation scheme is used to control the Switches of the VSC [4, 5]. The control functions will be written in Python. The reader is recommended to read Chap. 2 for a quick introduction to Python and follow it up with a few tutorials on the Internet or [1]. The chapter will attempt to make the process as descriptive as possible with tips to avoid errors. The reader is advised to refer to Chaps. 3 and 4 for details on the circuit component objects and the control function descriptors and variables.

5.2 Description of the Circuit

Figure 5.1 shows the circuit that will be finally simulated. The circuit in Fig. 5.1 is a shunt-connected VAR compensator connected to a three-phase system where a three-phase source feeds a three-phase load. Though this circuit is not overly complex for an experienced power engineer, it has sufficient complexity that it requires every facility provided by the circuit simulator to be utilized and therefore makes a good case study for a tutorial on using the simulator.

The circuit in Fig. 5.1 is divided into parts so as to be able to describe each part in detail. The circuit in Fig. 5.2 shows the three-phase voltage source Va, Vb, and Vc. The voltages are considered to be balanced sinusoids but can be considered to

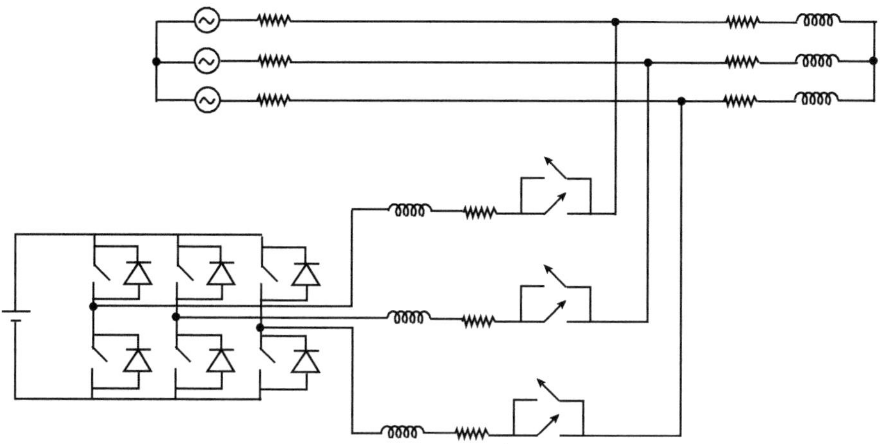

Fig. 5.1 Overall layout of the system with the VAR compensator

Fig. 5.2 Three-phase source

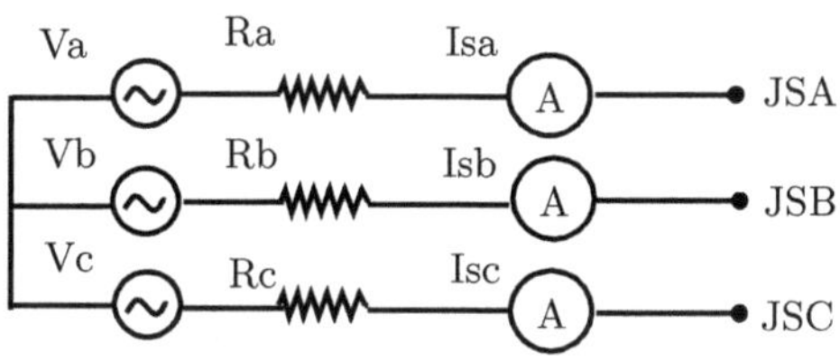

Fig. 5.3 Three-phase load

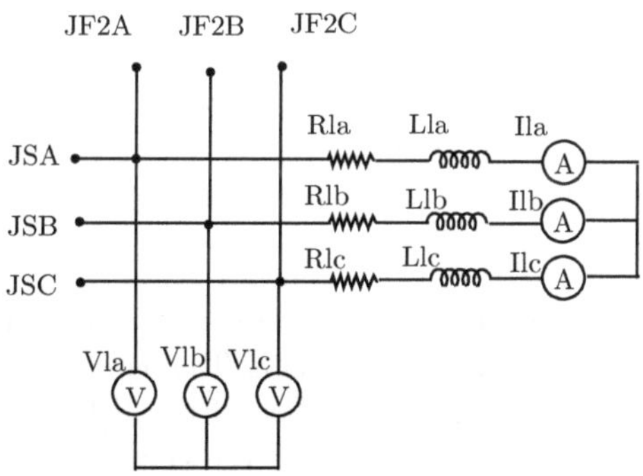

be unbalanced and with harmonics if so needed. A resistive three-phase feeder Ra, Rb, and Rc connects the source to the rest of the circuit. The source currents are measured by Ammeters Isa, Isb, and Isc. The three-phase source is connected to the rest of the circuit through jump labels JSA, JSB, and JSC.

The circuit in Fig. 5.3 shows the three-phase load. The load is considered to be linear and passive with only a resistance and an inductance. However, adding nonlinearity through Diode rectifiers and other inverters is always possible. The load current is measured by Ammeters. The voltages available at the load are measured by

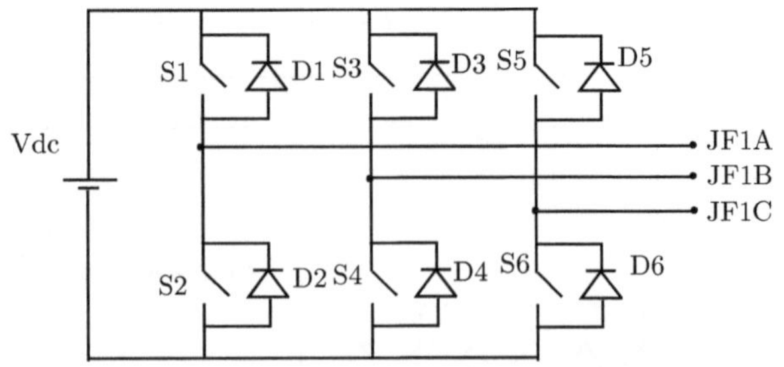

Fig. 5.4 Three-phase voltage source converter (VSC)

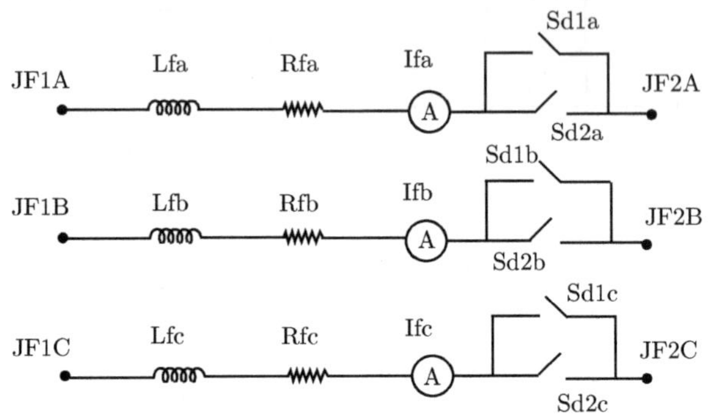

Fig. 5.5 Filter at the output of the VSC

Voltmeters. The load voltages and the load currents are used by the control algorithm to generate the references for the currents to be injected by the compensator. The circuit also shows connections that end with jump labels JF2A, JF2B, and JF2C. These jump labels connect the VAR compensator to the system. This entire part of the circuit is connected to the source through jump labels JSA, JSB, and JSC.

The circuit in Fig. 5.4 shows the three-phase voltage source converter. The converter is supplied by a dc voltage source Vdc. The nonlinear part of the converter is comprised of six Switches S1–S6. These Switches have antiparallel Diodes D1 to D6 associated with them to ensure that current through the inductors at the output of the converter freewheels. The output of the inverter is connected to the rest of the circuit through jump labels JF1A, JF1B, and JF1C.

The circuit in Fig. 5.5 shows the filter connected to the output of the converter. In this case, a simple inductor filter has been considered. Higher-order filters such as inductor–capacitor–inductor (L-C-L) filters at the output of the converter will result in better waveforms of the currents injected by the converter. However, this filter serves the purpose of demonstrating the performance of the converter. The filter has a bidirectional Switch connected in every phase. The purpose of this Switch is to be able to connect the VSC only after an initial grid synchronization has been performed so as to avoid excessive transients when the compensator is connected to the system. The currents produced by the converter are measured by the Ammeters Ifa, Ifb, and Ifc.

The circuit is split up into separate spreadsheets in the manner shown above. This facility is possible in the simulator where a circuit can be represented over multiple spreadsheets. As will be shown in the chapter, the circuit of Fig. 5.1 is developed in a modular manner. Moreover, it is much more convenient when the user wishes to change one part of the circuit. For example, in the above circuit, if the user wishes to replace the two-level voltage source converter with a three-level voltage source converter, only one spreadsheet needs to be changed and none of the others need to be altered. If the entire circuit is in one spreadsheet, it may have been necessary to move components around to accommodate the changes in one part of the circuit. Moreover, once a component such as the converter above has been created in a separate spreadsheet, it can be reused in any other circuit by changing only the jump labels.

5.3 Parameters of the Simulation and the Circuit

The simulator is launched with the command:

```
python circuit_solver.py
```

The first step—the simulation parameters in the spreadsheet circuit_inputs.csv. When a simulation is started from scratch, the spreadsheet circuit_inputs.csv has default values to help users with their simulations. Details of this can be found in Chap. 3. For the above circuit, the simulation parameters can be listed in Table 5.1. Below is the description of every row of the above table:

Table 5.1 Simulation parameters

Name of the circuit file	three_ph_source.csv	three_ph_load.csv	comp_inverter.csv	comp_filter.csv
Time duration of simulation	0.5			
Time step of the simulation	1.0e-6			
Time step of data storage	10.0e-6			
Name of the data storage file	ckt_output.dat			
Name of control files	pll.py	comp_reference.py	currcon_inverter.py	modulator.py
Split the output file?	Yes			
Length of time windows	0.25			

Row1 → Name of the circuit file: The simulation can have multiple spreadsheets containing parts of the circuit. Each spreadsheet as a .csv file will have to be specified in a separate cell in this row. The order of files is not important. It is necessary to specify the .csv extension of the files. In this simulation, the spreadsheets contain parts of the circuit shown above—three-phase source, three-phase load, converter, and filter.

Row2 → Time duration of simulation: This is time in seconds for which the simulation needs to be run. Time starts at 0 always. If the time is in minutes, it must be converted into seconds.

Row3 → Time step of the simulation: This is the time step of the ODE solver. An extremely critical parameter of the simulation, a large number could result in an unstable simulation. For most power electronics applications, 1 μs is typically sufficiently small. However, in some cases like resonant converters, this might have to be smaller. In general, it is advisable to reduce the time step of the simulation if it is unstable at the first choice.

Row4 → Time step of data storage: This is the time interval at which data is stored. It must be greater than or equal to the time step of simulation. The user needs to check on what the resolution of the stored data needs to be.

Row5 → Name of the data storage file: This can be any text file into which the data is stored. It should comply with file name rules relevant to the operating system in use.

Row6 → Name of the control files: These are the control files in the simulation. These are all Python files with a .py extension. Failure to specify the .py extension will cause an error. In this simulation, the control files are—phase-locked loop, current reference generator, converter current controller, and pulse width modulator.

Row7 → Split the output file: This is a facility for simulations that are large and the size of the data file can be expected to be excessively large. A large data file could be difficult to plot and may cause plotting software to either hang or crash. Specifying a Yes here will check the next row for the time window of each file.

Row8 → Length of time windows: If the answer to the previous row was a Yes, this number will be the window of the time interval in seconds for each output file. In this case, each output file will have 0.25 s of data and since the time duration of the simulation is 0.5 s, this implies there will be two output data files. These two output data files will have the same name as that specified but will be appended by the serial number of the file. So, in this case, the output files will be—ckt_output1.dat and ckt_output2.dat. In general, the format is (filename)index.(extension).

After providing the above simulation parameters to the simulator, the circuit components in all the spreadsheets will be read and objects will be created for all of them with default parameters. The simulator will then check if parameter spreadsheets exist for the circuit files in the directory. For example, for the circuit spreadsheet three_ph_source.csv, the simulator will look for a parameter spreadsheet three_ph_source_params.csv. If such a spreadsheet exists, the parameters of the circuit components in three_ph_source.csv will be read from this spreadsheet and the

Table 5.2 Three-phase source parameters

VoltageSource	Vsource1a	1C	Peak (Volts) = 169.70, frequency (Hertz) = 60.0, phase (degrees) = 0.0, dc offset = 0.0, positive polarity toward (cell) = 1D
VoltageSource	Vsource1b	11C	Peak (Volts) = 169.70, frequency (Hertz) = 60.0, phase (degrees) = −120.0, dc offset = 0.0, positive polarity toward (cell) = 11D
VoltageSource	Vsource1c	20C	Peak (Volts) = 169.70, frequency (Hertz) = 60.0, phase (degrees) = −240.0, dc offset = 0.0, positive polarity toward (cell) = 20D
Ammeter	Isource1a	1L	Positive polarity toward (cell) = 1M
Ammeter	Isource1b	11L	Positive polarity toward (cell) = 11M
Ammeter	Isource1c	20L	Positive polarity toward (cell) = 20M
Resistor	Rsource1a	1F	0.1
Resistor	Rsource1b	11F	0.1
Resistor	Rsource1c	20F	0.1

objects for the circuit components in this spreadsheet will be updated. If a parameter spreadsheet does not exist, the simulator will create a default parameter spreadsheet with the default parameters of the circuit components. In this manner, the user can edit this parameter spreadsheet instead of creating one from scratch. The parameters of each circuit spreadsheet will be listed in the following tables.

For the circuit spreadsheet three_ph_source.csv, the parameters of three_ph_source_param-s.csv are listed in Table 5.2. It should be noted that the fourth column of Table 5.2 contains all the parameters to make better use of space. The simulator will however need each parameter in a separate cell in the same row of the spreadsheet or will throw an error. It is recommended to modify the sample parameter spreadsheet created by the simulator by default or use sample spreadsheets from the Web site. Creating an empty spreadsheet and entering the following data into it will result in errors and is not recommended. The voltage sources are chosen to have a peak of 169.7 V which corresponds to the RMS value of 120 V. The phase b voltage lags behind the phase a voltage by 120°, while the phase c voltage lags behind the phase a voltage by 240°. Therefore, the three-phase voltage source is a balanced sinusoidal source.

For the circuit spreadsheet three_ph_load.csv, the parameters of three_ph_load_params.csv are listed in Table 5.3. The load has been chosen to have a reactance of 7.54 Ω at 60 Hz. Therefore, the load is predominantly inductive and has a poor power factor since it draws a large amount of reactive power. This provides an incentive to perform VAR compensation. A note about the Voltmeters. The voltage range of the Voltmeters has been chosen to be 600 V. There is only one factor to be taken into consideration while choosing the voltage range. The Voltmeter is a component that will draw a negligible current at rated voltage which in this circuit is 120 V. To ensure that the current drawn by the Voltmeter is never non-negligible, it is advisable

Table 5.3 Three-phase load parameters

Inductor	Lload1a	1J	0.02
Inductor	Lload1b	9J	0.02
Inductor	Lload1c	17J	0.02
Resistor	Rload1a	1G	4
Resistor	Rload1b	9G	4
Resistor	Rload1c	17G	4
Ammeter	Iload1a	1M	Positive polarity toward (cell) = 1N
Ammeter	Iload1b	9M	Positive polarity toward (cell) = 9N
Ammeter	Iload1c	17M	Positive polarity toward (cell) = 17N
Voltmeter	Vload1a	36G	Rated voltage level to be measured = 600.0, positive polarity toward (cell) = 35G
Voltmeter	Vload1b	36J	Rated voltage level to be measured = 600.0, positive polarity toward (cell) = 35J
Voltmeter	Vload1c	36M	Rated voltage level to be measured = 600.0, positive polarity toward (cell) = 35M

to choose a voltage range that is a multiple of the rated voltage. This will ensure that even in the worst case, the Voltmeter will always be a stiff branch. The voltage range of the Voltmeter could have been chosen to be 1000 V in the above case if the user wishes to be safe. Choosing anything close to 120 V is however not recommended.

For the circuit spreadsheet comp_inverter.csv, the parameters of comp_inverter_params.csv are listed in Tables 5.4 and 5.5. There is only one component for VoltageSource. If a dc voltage is needed, the peak which is the ac peak and the frequency should be made zero while the dc offset should be set to the desired dc voltage. With the Switches and Diodes, caution needs to be exercised while specifying the polarity. As with the Voltmeter, the voltage range of the Switches and Diodes is set to 600 V to ensure that for the worst voltage, the devices will draw a negligible current when they are in their off state. The branch having the dc voltage source has a resistance of 0.1 Ω as it is illegal to have a branch in the circuit with zero resistance. The resistance can be reduced to a smaller value if it is found to disrupt the operation of the converter. The Switches are controllable components with a gate signal that can be specified under "Name of control signal." This has the default value of "Control" when an object corresponding to a Switch is created in the simulator. However, this should be changed to a unique value that links to the particular Switch. The control signals for the Switches S1–S6 are named S1compinvgate to S6compinvgate. It is recommended to make the names of the control signal verbose and descriptive as the above part of the circuit could be used in a larger circuit with other power electronic converters at a later stage.

For the circuit spreadsheet comp_filter.csv, the parameters of comp_filter_params.csv are listed in Table 5.6. The filter chosen for the compensator is a 1 mH inductor. The voltage range for the disconnect Switches is also chosen as 600 V like

Table 5.4 Three-phase VSC parameters

VoltageSource	invdcbus	15A	Peak (Volts) = 0.0, frequency (Hertz) = 0.0, phase (degrees) = 0.0, dc offset = 400.0, positive polarity toward (cell) = 14A
Resistor	invdcbus	8A	0.1
Switch	S1compinv	8H	Voltage level (V) = 600.0, negative polarity toward (cell) = 9H, name of control signal = S1compinvgate
Switch	S2compinv	22H	Voltage level (V) = 600.0, negative polarity toward (cell) = 23H, name of control signal = S2compinvgate
Switch	S3compinv	8O	Voltage level (V) = 600.0, negative polarity toward (cell) = 9O, name of control signal = S3compinvgate
Switch	S4compinv	22O	Voltage level (V) = 600.0, negative polarity toward (cell) = 23O, name of control signal = S4compinvgate
Switch	S5compinv	8T	Voltage level (V) = 600.0, negative polarity toward (cell) = 9T, name of control signal = S5compinvgate
Switch	S6compinv	22T	Voltage level (V) = 600.0, negative polarity toward (cell) = 23T, name of control signal = S6compinvgate

Table 5.5 Three-phase VSC parameters

Diode	D1compinv	8J	Voltage level (V) = 600.0, cathode polarity toward (cell) = 7J
Diode	D2compinv	22J	Voltage level (V) = 600.0, cathode polarity toward (cell) = 21J
Diode	D3compinv	8Q	Voltage level (V) = 600.0, cathode polarity toward (cell) = 7Q
Diode	D4compinv	22Q	Voltage level (V) = 600.0, cathode polarity toward (cell) = 21Q
Diode	D5compinv	8V	Voltage level (V) = 600.0, cathode polarity toward (cell) = 7V
Diode	D6compinv	22V	Voltage level (V) = 600.0, cathode polarity toward (cell) = 21V

before. It is a safe practice to choose the voltage rating of stiff elements and nonlinear devices to be the same as the simulator deals with resistances of the same magnitude. Caution should be exercised when choosing the polarity of the Switches—the antiparallel Diodes in a phase should have opposite polarities.

A few observations to be made about the parameters in the above tables. As stated before, the simulator allows users to split their circuit over multiple schematics. In all the schematics, the cell position of the components and also their direction if any are specified only with respect to that spreadsheet. The simulator will keep track of which component appears in which spreadsheet. For a user, each spreadsheet

Table 5.6 Filter parameters

Inductor	Lcompa	1J	0.001
Inductor	Lcompb	9J	0.001
Inductor	Lcompc	17J	0.001
Resistor	Rcompa	1H	0.1
Resistor	Rcompb	9H	0.1
Resistor	Rcompc	17H	0.1
Ammeter	Icompa	1M	Positive polarity toward (cell) = 1N
Ammeter	Icompb	9M	Positive polarity toward (cell) = 9N
Ammeter	Icompc	17M	Positive polarity toward (cell) = 17N
Switch	Disconnect1a	1R	Voltage level (V) = 600.0, negative polarity toward (cell) = 1Q, name of control signal = Disconnect1a
Switch	Disconnect2a	4R	Voltage level (V) = 600.0, negative polarity toward (cell) = 4S, name of control signal = Disconnect2a
Switch	Disconnect1b	9R	Voltage level (V) = 600.0, negative polarity toward (cell) = 9Q, name of control signal = Disconnect1b
Switch	Disconnect2b	12R	Voltage level (V) = 600.0, negative polarity toward (cell) = 12S, name of control signal = Disconnect2b
Switch	Disconnect1c	17R	Voltage level (V) = 600.0, negative polarity toward (cell) = 17Q, name of control signal = Disconnect1c
Switch	Disconnect2c	20R	Voltage level (V) = 600.0, negative polarity toward (cell) = 17S, name of control signal = Disconnect2c

becomes a separate entity. The only exception being that the name of a component has to be unique in all the spreadsheets. It is strongly recommended that a user tries out variations of the above example.

5.4 First Stage in Control Development—Grid Synchronization

This circuit has been chosen as an example to demonstrate how the circuit along with the associated control can be developed in stages. Doing so may appear to increase development time but reduces errors and therefore ends up reducing the total time required to generate the complete simulation model. The first stage in developing the simulation model is connecting the three-phase source to the three-phase load without the compensator. Connecting the three-phase load to the three-phase source is fairly simple. Let us use this simple case to generate the grid synchronization algorithm—the phase-locked loop (PLL). For any grid-connected system, the PLL is a fundamental component. This is because almost every ac source will experience some change in one of its parameters—the voltage magnitude, frequency, or phase

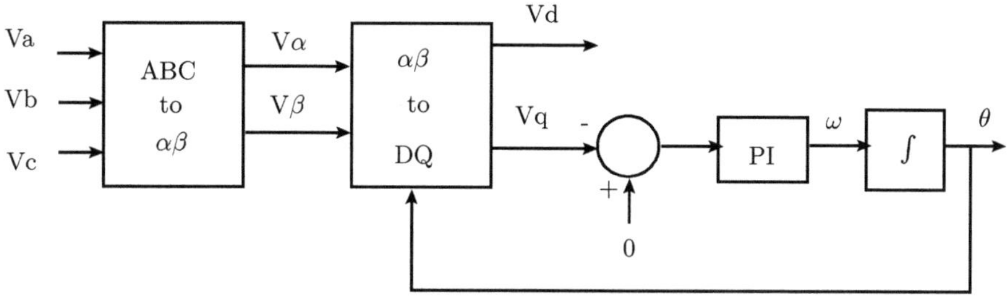

Fig. 5.6 Phase-locked loop (PLL)

angle. To assume the grid frequency to be constant will cause any current control algorithm to fail.

The simplest method of designing control for a simulation is to begin with the control function. After a preliminary control function has been developed, the control descriptor can be finalized. In case of the PLL, the control algorithm can be described by the block diagram of Fig. 5.6 [2]. The inputs to the PLL are the three measured voltages Va, Vb, and Vc which in this simulation example are the voltages measured at the load. These three voltages are converted into another frame of reference called the alpha–beta frame of reference. The voltages V_α and V_β are expressed as:

```
Valpha = Va − 0.5Vb − 0.5Vc
Vbeta = sqrt(3)*(Vb − Vc)/2
```

In this simulation example, the circuit is a three-wire circuit without a neutral wire. Therefore, such a circuit does not have a zero-sequence component in either the currents or the voltages. Valpha and Vbeta are sufficient to express the voltages of the system. If there was a neutral wire, a zero-sequence component would also be present. The voltages Valpha and Vbeta are transformed to another frame of reference called the D-Q frame of reference. The D-Q frame of reference is a rotating frame of reference. The voltages Vd and Vq are:

```
Vd = Valpha*cos(wt) + Vbeta*sin(wt)
Vq = −Vbeta*cos(wt) + Valpha*sin(wt)
```

In the above equations, w(ω) is the angular frequency of rotation of the frame of reference. If the frequency of rotation w is equal to the angular frequency of the voltages Va, Vb, and Vc, the PLL is said to have locked on to the grid. The PLL uses a proportional–integral (PI) controller to generate an angular frequency w such that Vq is forced to zero while Vd remains nonzero which is one of the many possible equilibrium states of the PLL. This angular frequency produced by the PI controller is fed back to the alpha–beta to D-Q transformation block to form a closed-loop synchronization control as shown in Fig. 5.6.

To test the PLL, we need only connect the three-phase source to the three-phase load. The converter and the filter are not necessary. Therefore, circuit_inputs.csv can be modified with the parameters listed in Table 5.7. The base code of the PLL can be written as the following block:

Table 5.7 Simulation parameters

Name of the circuit file	three_ph_source.csv	three_ph_load.csv
Time duration of simulation	0.5	
Time step of the simulation	1.0e-6	
Time step of data storage	10.0e-6	
Name of the data storage file	ckt_output.dat	
Name of control files	pll.py	
Split the output file?	Yes	
Length of time windows	0.25	

```
volt_alpha = math.sqrt(2.0/3.0)*(volt_a - 0.5*volt_b - \
                        0.5*volt_c)
volt_beta = math.sqrt(1.0/2.0)*(volt_b - volt_c)
volt_d = volt_alpha*math.cos(ph_angle) + \
              volt_beta*math.sin(ph_angle)
volt_q = -volt_alpha*math.sin(ph_angle) + \
              volt_beta*math.cos(ph_angle)
pll_err = volt_q
pll_int += pll_err*dt_sample
omega = nom_omega + pll_kp*pll_err + pll_ki*pll_int
ph_angle += omega*dt_sample
if (ph_angle > 2*math.pi):
        ph_angle = ph_angle - 2*math.pi
```

The first four expressions are the equations written before. The q component of the transformed voltage volt_q is considered to be the PLL error and fed to the integrator of the PI controller with proportional gain pll_kp and integral gain pll_ki. Since the angular frequency is not a completely random number but a nominal value exists for every system, it would decrease the control settling time if the angular frequency omega were generated around this nominal frequency nom_omega rather than starting from zero. This angular frequency omega is then integrated to produce the phase angle of the system ph_angle. Typically, the phase angle changes over an interval of 2π as any sinusoid has a period of 2π radians. However, the output of the integrator can increase indefinitely. To ensure that the ph_angle is limited, the next check statement ensures that phase angle remains limited to 0 to 2π.

If the above control block is placed in a control file in the above form, it will run at every simulation time step which is 1 µs. It is completely unnecessary to run a PLL algorithm at 1 MHz when the grid is at 60 Hz. It would be quite sufficient to run the PLL control algorithm every 100 µs or at 10 kHz as this will provide sufficient accuracy in the phase angle synchronization and at the same time will not be a computational burden. The addition will be a time event—let us call it t1. The PLL control block can be written as:

```
dt_sample = 100.0e-6
pll_kp = 0.2
pll_ki = 4.5
nom_omega = 370.0
if  t_clock>=t1:
        <control block above>
        t1 += dt_sample
```

As described in Chap. 4, every control function is provided with a reserved variable t_clock which is the current time instant of the simulation. This can be used for synchronizing the control function as shown above. Only when t_clock exceeds or is equal to the time event t1, does the control block execute. After execution, the time event is incremented by the sampling time of the control function, which in this case is 100 μs as described above. For further details on sampling of control functions, refer to Chap. 4. Besides this, the PI controller gains are defined at the beginning of the control function before the time block. These gains can be obtained in a number of different ways—through frequency domain analysis or by simple trial-and-error method. The nominal angular frequency is defined to be 370 rad/s. For a 60-Hz system, the angular frequency is 377 rad/s. Typically, the frequency of a power system can change by 1–2%. Therefore, the nominal angular frequency could be chosen differently for different systems.

The last component of the control function is the interface to the rest of the simulation. Before elaborating on this, let us examine the nature of the variables in the control function above. There are so far two types of variables. The first are the local variables. These are dt_sample, pll_kp, pll_ki, and nom_omega. These are defined at the beginning of the control file and therefore will be created and assigned values every time the control function is called. These objects will be destroyed once the function is executed. These variables could also have been declared as static variables. The choice is completely to the user. For details on static variables, refer to Chap. 4. It should be noted that if a user does not wish to declare a variable as static, it should be declared at the head of the control function to ensure that it always exists at every evaluation of the control function. A static variable is created and comes into existence when the simulation begins. However, if a variable is created within the time block, it is possible that when the time block does not execute, the variable will never be created. In the event that this variable is accessed outside the time block, an error will be generated as an object is being accessed that was never created. Therefore, every variable must either be assigned at the head of the control function or must be declared as a static variable to ensure that no error is thrown while the control function is executed during the simulation.

With the nature of local variables and static variables discussed, let us get back to the variables used to interface the control function to the rest of the simulation. These variables are called "VariableStorage." For details on these, refer to Chap. 4. VariableStorage types were originally designed for debugging control functions. However, in order to have cascaded control functions where a control function does not directly change the circuit through a electrical component, it is necessary to have a variable type that can be accessed by another control function and if necessary be written to the output data file for plotting. This is specifically the case with the PLL function above. It takes as input the measured load voltages but does not produce any signal that changes the state of the circuit. On the contrary, the circuit does not have any controllable components such as Switches at this stage. With reference to the block diagram of the PLL, the output of the PLL that will be used by the external simulation is the generated phase angle ph_angle. This provides the angle information of the grid and will be used by other control functions as will be described later. Besides this, for the purpose of examining the working of a PLL, let us also plot the generated angular frequency omega and the d and q components of voltages that the PLL used to lock on the measured grid voltages. The following block shows the output interface of the PLL control function:

```
pll_phase_angle = ph_angle
pll_omega = omega
pll_volt_d = volt_d
pll_volt_q = volt_q
```

The "pll_" is added to these output variables to differentiate them from the output variables of other control functions. To elaborate on this, the user should refer to Chap. 4 on VariableStorage type. Any variable of type VariableStorage is common to all control functions of the simulation. This implies the variables pll_phase_angle, pll_omega, pll_volt_d, and pll_volt_q will be available in all other control functions as will be added later. Additionally, any change made to these variables in other control functions will be reflected back. Therefore, these variables can be viewed as global variables in a simulation. Despite an extremely powerful concept, it also opens the doors to bugs that are very hard to fix as variables are not constrained to a control function. It is therefore advisable to restrict the use of the VariableStorage type to the minimum required to interface functions and debug variables. Delete any additional variables of this type once they are no longer required and the control has been tested. In this manner, if these variables have been accidentally changed in another control function, an error will be generated that an object is being referenced that has not been created.

With the above description of the different types of variables, the descriptor file for the control function can now be edited. For details on the descriptor file of a control function, refer to Chap. 4. Every control function in a simulation has a descriptor file. The control file pll.py will have a descriptor pll_desc.csv. When the simulator is launched, it will check if a control function has a descriptor. If a descriptor exists, the user will be asked to check if it has been updated. If no descriptor exists, the simulator will create a blank descriptor which the user can edit. The descriptor pll_desc.csv is listed in Tables 5.8 and 5.9.

Table 5.8 Descriptor pll_desc.csv for pll.py

Input	Element name in circuit spreadsheet = Voltmeter_Vload1a, desired variable name in control code = volt_a
Input	Element name in circuit spreadsheet = Voltmeter_Vload1b, desired variable name in control code = volt_b
Input	Element name in circuit spreadsheet = Voltmeter_Vload1c, desired variable name in control code = volt_c
TimeEvent	Desired variable name in control code = t1, first time event = 0.0
VariableStorage	Desired variable name in control code = pll_phase_angle, initial value of variable = 0.0, plot variable in output file = yes
VariableStorage	Desired variable name in control code = pll_omega, initial value of variable = 0.0, plot variable in output file = yes
VariableStorage	Desired variable name in control code = pll_volt_d, initial value of variable = 0.0, plot variable in output file = yes
VariableStorage	Desired variable name in control code = pll_volt_q, initial value of variable = 0.0, plot variable in output file = yes

Table 5.9 Descriptor pll_desc.csv for pll.py

StaticVariable	Desired variable name in control code = ph_angle, initial value of variable = 0.0
StaticVariable	Desired variable name in control code = volt_alpha, initial value of variable = 0.0
StaticVariable	Desired variable name in control code = volt_beta, initial value of variable = 0.0
StaticVariable	Desired variable name in control code = volt_d, initial value of variable = 0.0
StaticVariable	Desired variable name in control code = volt_q, initial value of variable = 0.0
StaticVariable	Desired variable name in control code = pll_d, initial value of variable = 0.0
StaticVariable	Desired variable name in control code = pll_q, initial value of variable = 0.0
StaticVariable	Desired variable name in control code = pll_err, initial value of variable = 0.0
StaticVariable	Desired variable name in control code = pll_int, initial value of variable = 0.0
StaticVariable	Desired variable name in control code = omega, initial value of variable = 0.0

Let us simulate this system and look at the simulation results. In order to plot the variables, a plotting software like Gnuplot is needed. Any plotting software will do as long as it can plot one column of data against another as each data item is stored as a column in the output file separated by a space. For example, for the above circuit, the output data will be plotted in the following manner:

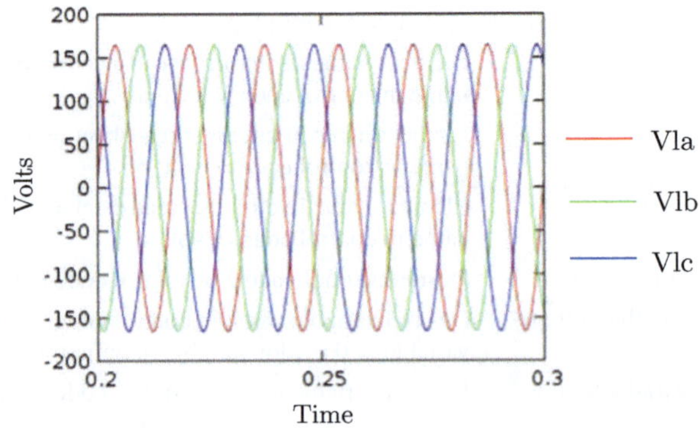

Fig. 5.7 Three-phase voltages

```
*************************************************
Output file is in ckt_output.dat and the columns are in the
following sequence
1 => Time
Meters are in the following sequence:
2   =>  Ammeter_Isource1a at  1L
3   =>  Ammeter_Isource1b at  11L
4   =>  Ammeter_Isource1c at  20L
5   =>  Ammeter_Iload1a at  1M
6   =>  Ammeter_Iload1b at  9M
7   =>  Ammeter_Iload1c at  17M
8   =>  Voltmeter_Vload1a at  36G
9   =>  Voltmeter_Vload1b at  36J
10  =>  Voltmeter_Vload1c at  36M

Control variables to be plotted are in the following sequence
11  =>  pll_omega
12  =>  pll_volt_q
13  =>  pll_cont_int
14  =>  pll_volt_d
15  =>  pll_phase_angle
*************************************************
```

For any simulation, column 1 will always be the simulation time instant. The next set of data items will be the measured outputs of meters. The last set of data items will be the variables stored by the control functions.

To begin with, let us examine the three-phase load voltages in Fig. 5.7. Since both the source and the load are balanced and linear, the load voltages are balanced and sinusoidal. The next two plots in Figs. 5.8 and 5.9 show the load currents and the phase currents which in this circuit are equal. Before examining the PLL performance, let us examine the power factor of the load by plotting phase a voltage scaled down by a factor of 5 (Va/5) and the phase a load current in Fig. 5.10. Figure 5.10 shows the phase a load current lagging behind the phase a load voltage by a fairly large angle. This indicates that the load is drawing a large amount of reactive power from

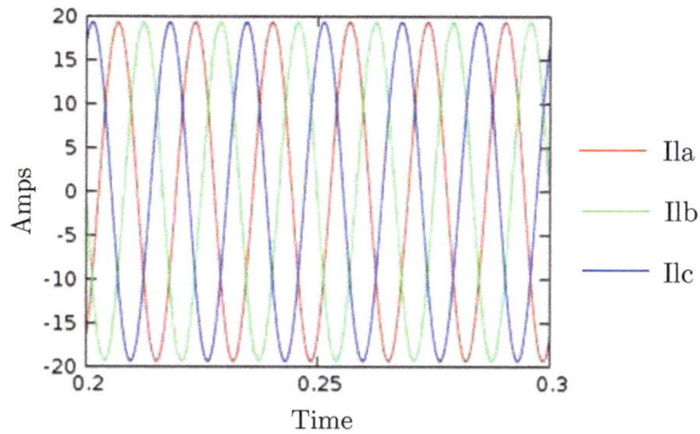

Fig. 5.8 Load currents

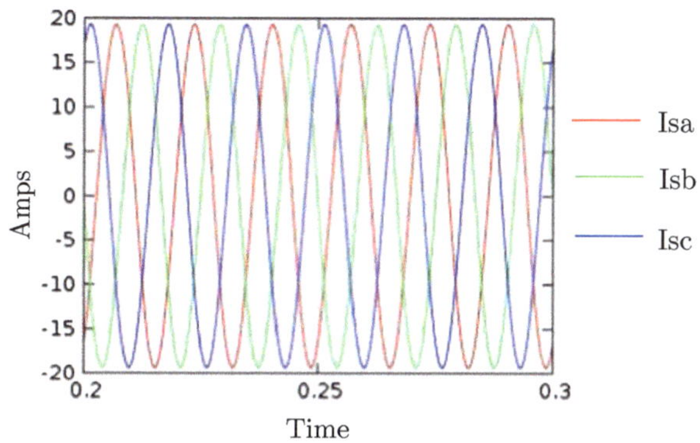

Fig. 5.9 Source currents

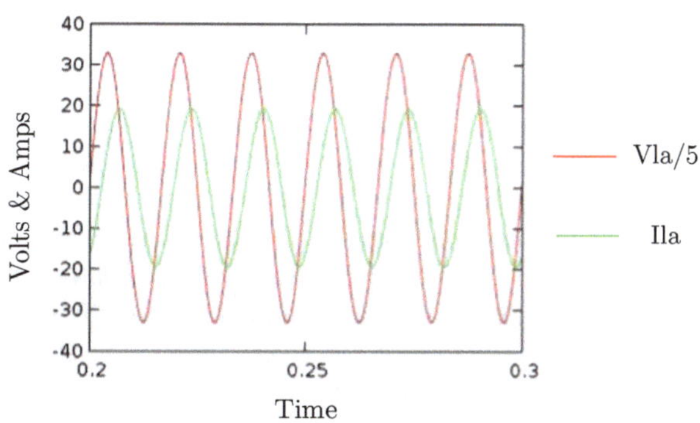

Fig. 5.10 Power factor of the load

the source. The later simulation results will show the VAR compensator reduces the burden on the source by supplying this current.

Figure 5.11 shows how the output angular frequency of the PLL changes due to the closed-loop control scheme. The angular frequency starts at the nominal frequency defined in the control function to be 370 rad/s to gradually increase and settle at 377 rad/s. There is a slight overshoot before settling which can be minimized by tuning the PI controller gains. The plot in Fig. 5.12 shows the tracking performance of the PLL. The PLL as shown in the block diagram forces the q component of

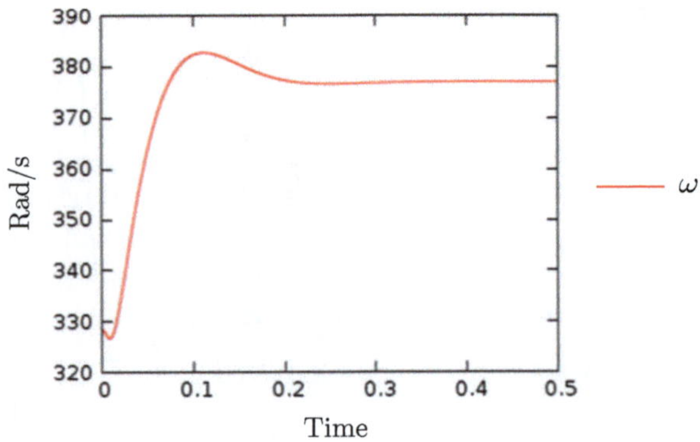

Fig. 5.11 Frequency output of PLL

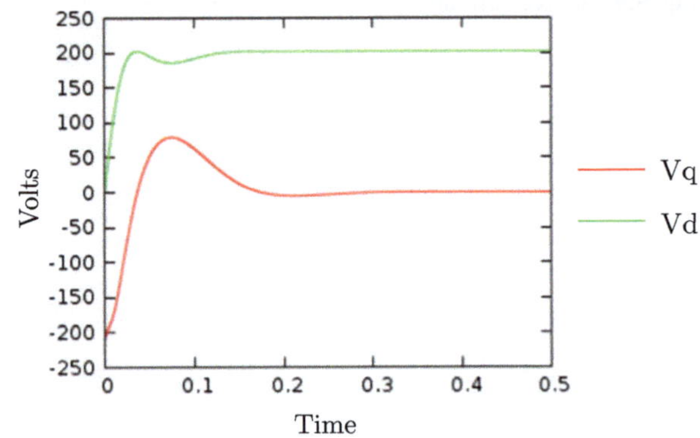

Fig. 5.12 Voltage output of PLL

the transformed voltage in the synchronous reference frame to zero. This ensures that the synchronous reference frame is rotating at the same angular frequency as the grid. The plot in Fig. 5.13 shows the synchronization effect of the PLL by plotting the phase angle with the phase a grid voltage by scaling the grid voltage down by a factor of 10. As can be seen from the plot, the phase angle generated will change from 0 rad (0°) to 2 pi rad (360°) as the phase a voltage changes from positive maximum to zero to negative maximum to zero and back to positive maximum. By mere observation, calculating the cosine of the phase angle generated will produce a template of the phase a voltage with unity magnitude.

5.5 Second Stage in Control Development—Current Reference Generation and Closed-Loop Current Control

The above plots show that the tracking performance of the PLL is adequate. Improvements can be made by tuning the PI controller parameters or the nominal angular frequency. However, for this particular circuit, the above PLL is sufficient in terms

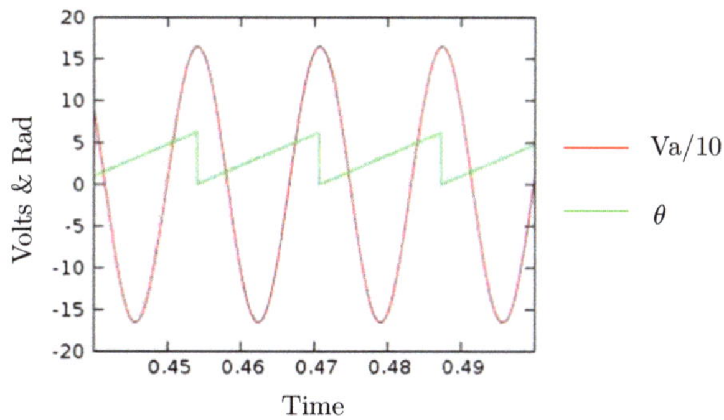

Fig. 5.13 Phase angle output of PLL

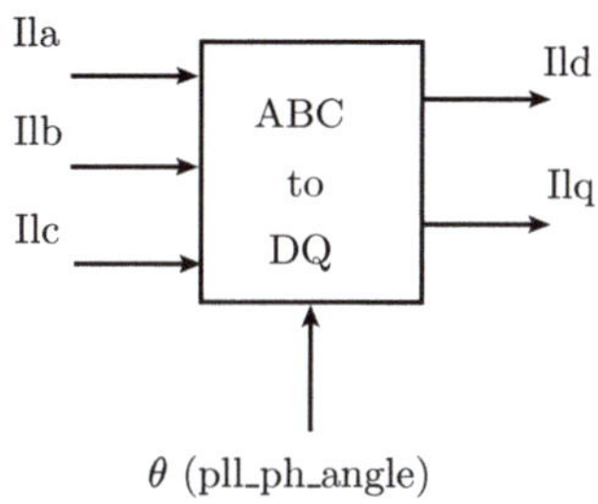

Fig. 5.14 Compensator reference current generation

of reasonably fast response and good steady-state performance. Therefore, we shall continue with simulating the rest of the circuit. Before including the VAR compensator, let us examine the method of generating the references for the currents to be supplied by the compensator. The transformation of three-phase variables to the synchronously rotating d-q reference frame has a quality that makes it convenient for generating current references for a VAR compensator. In the above PLL, the angular frequency of the rotating d-q reference frame is changed until the q component of voltage is zero. This state results in angular frequency of the rotating frame to be equal to the angular frequency of the system and also results in the d component of the voltage to be aligned with the phase a load voltage while the q component is in quadrature with the phase a voltage. The similar concept of in-phase and quadrature components can be applied to the load currents. By transforming the three-phase load currents to the d-q reference frame, the d component of the load current will be in phase with the phase a load voltage, while q component of the load current will be in quadrature with the phase a load voltage. The concept is described by the block diagram in Fig. 5.14.

The control code for the generation of current references for the VAR compensator is present in comp_reference.py. The block of code that converts the three-phase currents to d-q components is below:

```
curr_alpha = math.sqrt(2.0/3.0)*(curr_a - 0.5*curr_b \
                 - 0.5*curr_c)
curr_beta = math.sqrt(2.0/3.0)*math.sqrt(3)*(curr_b \
                 - curr_c)/2.0
```

```
    curr_d  = curr_alpha*math.cos(pll_phase_angle) + \
                        curr_beta*math.sin(pll_phase_angle)
    curr_q  = -curr_alpha*math.sin(pll_phase_angle) + \
                        curr_beta*math.cos(pll_phase_angle)
```

As with the case of the load voltages in the PLL, the currents are first transformed to the stationary alpha–beta reference frame and then to the rotating d-q reference frame. It should be noted that the variable pll_phase_angle has been defined in pll_desc.csv and generated in pll.py. By using VariableStorage types, an interface between two control functions can be formed. It should be added that using pll_phase_angle in the above manner is acceptable. However, if any manipulations are to be made with respect to this variable, a static variable should be defined and pll_phase_angle should be assigned to it. This is due to the fact that any changes made to pll_phase_angle in this control function could disrupt the PLL performance.

The next part of the control function is the timing of the control function. The reference generation can be at the same frequency as the PLL as this function has to be significantly faster than the fundamental frequency of the system. Therefore, the time period of the function is set to be 100 μs.

```
dt_sample = 100.0e-6
if t_clock>=t1:
        <Above code here>
        t1 += dt_sample
```

As with the PLL, we use the in-built variable t_clock to check the time instant of the simulation. The variable t1 will be defined in the descriptor file as a TimeEvent variable. This variable is updated at the end of the block to mark the next time instant the reference generation algorithm must run.

As with the case of the PLL, the output of the reference generation function must interface with other control functions. Therefore, d and q components of the load current must be output to the current control block which will be described next. Since an interface between control functions is to be formed, VariableStorage type is used. The following two variables will provide the d-q components of load currents to the rest of the simulation:

```
comp_curr_d = curr_d
comp_curr_q = curr_q
```

The entire control function is fairly simple and can be listed here in its entirety:

```
dt_sample = 100.0e-6
if t_clock>=t1:
        curr_alpha = math.sqrt(2.0/3.0)*(curr_a - 0.5*curr_b \
                        - 0.5*curr_c)
        curr_beta = math.sqrt(2.0/3.0)*math.sqrt(3)*(curr_b \
                        - curr_c)/2.0
        curr_d = curr_alpha*math.cos(pll_phase_angle) + \
                        curr_beta*math.sin(pll_phase_angle)
        curr_q = -curr_alpha*math.sin(pll_phase_angle) + \
```

Table 5.10 Simulation parameters

Name of the circuit file	three_ph_source.csv	three_ph_load.csv
Time duration of simulation	0.5	
Time step of the simulation	1.0e-6	
Time step of data storage	10.0e-6	
Name of the data storage file	ckt_output.dat	
Name of control files	pll.py	comp_reference.py
Split the output file?	Yes	
Length of time windows	0.25	

```
                    curr_beta*math.cos(pll_phase_angle)
        t1 += dt_sample
comp_curr_d = curr_d
comp_curr_q = curr_q
```

The circuit_inputs.csv file for this run of the simulation can be changed as listed in Table 5.10. The circuit still contains only the three-phase source and the three-phase load. Only an additional control function comp_reference.py has been added. The descriptor file can be edited as listed in Table 5.11. Now that everything is in place, we can launch the simulator. It should be noted that if the user wishes to experiment with the descriptor file or the control function, it is advisable to delete these files from the directory and let the simulator populate comp_reference_desc.csv with default values. The user can then edit this file and examine how the simulator generates errors for objects that do not exist because these are not defined as StaticVariables. Upon launching the simulator and confirming the contents of the parameters and the descriptors, the data written into the output data file ckt_output.dat will be as follows:

```
*************************************************
Output file is in ckt_output.dat and the columns are in the
following sequence
1 => Time
Meters are in the following sequence:
2  =>  Ammeter_Isource1a at  1L
3  =>  Ammeter_Isource1b at  11L
4  =>  Ammeter_Isource1c at  20L
5  =>  Ammeter_Iload1a at  1M
6  =>  Ammeter_Iload1b at  9M
7  =>  Ammeter_Iload1c at  17M
8  =>  Voltmeter_Vload1a at  36G
9  =>  Voltmeter_Vload1b at  36J
10  =>  Voltmeter_Vload1c at  36M

Control variables to be plotted are in the following sequence
11  =>  pll_volt_d
12  =>  pll_cont_int
13  =>  comp_curr_d
14  =>  pll_phase_angle
```

Table 5.11 Descriptor comp_refernce_desc.csv for comp_refernce.py

Input	Element name in circuit spreadsheet = Ammeter_Iload1a, desired variable name in control code = curr_a
Input	Element name in circuit spreadsheet = Ammeter_Iload1b, desired variable name in control code = curr_b
Input	Element name in circuit spreadsheet = Ammeter_Iload1c, desired variable name in control code = curr_c
StaticVariable	Desired variable name in control code = curr_alpha, initial value of variable = 0.0
StaticVariable	Desired variable name in control code = curr_beta, initial value of variable = 0.0
StaticVariable	Desired variable name in control code = curr_d, initial value of variable = 0.0
StaticVariable	Desired variable name in control code = curr_q, initial value of variable = 0.0
TimeEvent	Desired variable name in control code = t1, first time event = 0.0
VariableStorage	Desired variable name in control code = comp_curr_d, initial value of variable = 0.0, plot variable in output file = yes
VariableStorage	Desired variable name in control code = comp_curr_q, initial value of variable = 0.0, plot variable in output file = yes

```
15  =>  pll_omega
16  =>  pll_volt_q
17  =>  comp_curr_q
*************************************************
```

As can be seen, the objects of type VariableStorage—comp_curr_d and comp_curr_q—have been included in the list of control variables to be written to the data file. Figure 5.15 shows the plot of these two variables. A couple of observations about this plot. From the parameters of the load, it is apparent that the load will draw a larger reactive power than active power. This can be observed by the green waveform which is the q component of the load current with respect to the red waveform which is the d component of the load current. Furthermore, due to the inductive nature of the load, the reactive power drawn by the load is lagging in nature, and therefore, the q component is negative. If the load had been capacitive instead, the reactive power drawn by the load would be leading and the q component of load current would be positive.

In this example, a fairly simple case has been chosen for current reference generation. The control function simply provides the d and q components of the load current. In most practical cases, a VAR compensator will be able to provide only a fraction of the reactive power demanded by a load. Moreover, the objective of VAR compensation is to regulate the voltage at the point of coupling through reactive power control. Therefore, in practice, a VAR compensator will need to process a much more complicated algorithm than merely computing the d and q components of the load current. In this example, it has been assumed that the VAR compensator has the capability to completely supply the q component of the current. For this

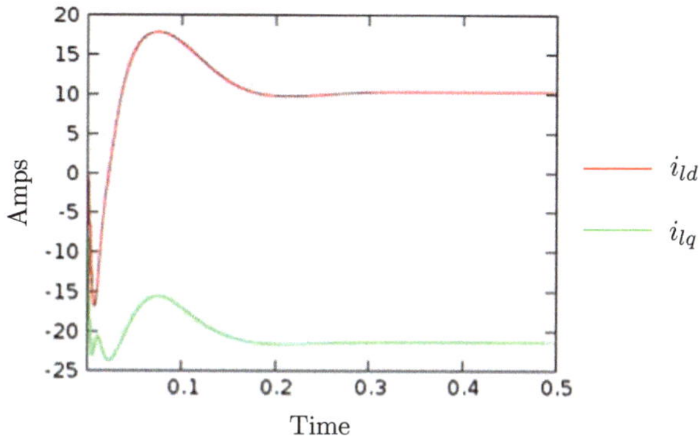

Fig. 5.15 Compensator reference currents

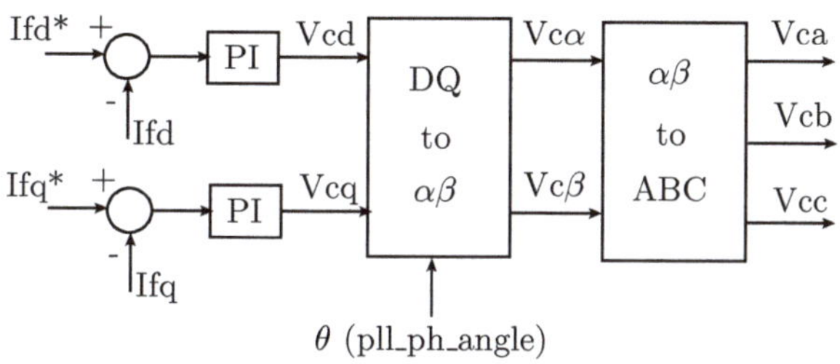

Fig. 5.16 Current control block diagram

reason, the above comp_reference.py is sufficient. The next step will be to describe the current control algorithm.

The block diagram of Fig. 5.16 shows the closed-loop current control algorithm [4]. Ifd* and Ifq* are the references for the d and q components of compensator current. Ifd and Ifq are the actual d and q components of the measured compensator current. In this example, Ifq* = Ilq but Ifd* = 0 as the compensator does not supply active power. Moreover, we will use a VSC that has a dc voltage source rather than a dc capacitor. In case of a dc capacitor, Ifd* will have a negative value as the compensator will need to draw a small amount of active power to maintain the voltage on the dc bus of the VSC. The objective of the control scheme is to ensure that the d and q components of the compensator current track the references that are based on the d and q components of the load current.

In Fig. 5.16, the output of the PI controllers are voltages Vcd and Vcq. Before simulating the VSC, let us consider the compensator to be made up of controllable voltage sources. This will allow us to tune the PI gains of the current controller which can then be transferred to the final simulation with the VSC. The outputs of the PI controller Vcd and Vcq can then be converted to three-phase voltages that are fed to the controllable voltage source. The circuit diagram is shown in Fig. 5.17. The above circuit is a combination of comp_inverter.csv and comp_filter.csv. The circuit is called comp_sources.csv since the compensator is comprised of controllable voltage sources. The parameters of the simulation are listed in Table 5.12, while the parameters of the circuit in Fig. 5.17 are listed in Tables 5.13 and 5.14. The

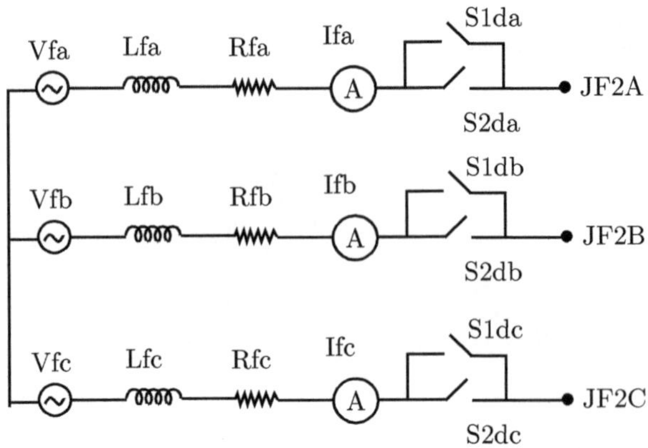

Fig. 5.17 Compensator formed of controllable voltage sources

Table 5.12 Simulation parameters

Name of the circuit file	three_ph_source.csv, three_ph_load.csv, comp_s-ources.csv
Time duration of simulation	0.5
Time step of the simulation	1.0e-6
Time step of data storage	10.0e-6
Name of the data storage file	ckt_output.dat
Name of control files	pll.py, comp_reference.py, currcon_sources.py
Split the output file?	Yes
Length of time windows	0.25

controllable elements in this circuit are the ControllableVoltageSources Vcompa, Vcompb, and Vcompc. The current controller will generate voltages for these three controllable sources such that the compensator will supply currents according to the references Ifd* and Ifq*.

Let us now examine the control code for generating voltages for the controllable sources. Below is the main control block:

```
curr_ref_d = 0.0
curr_ref_q = comp_curr_q

curr_alpha = math.sqrt(2.0/3.0)*(curr_a - 0.5*curr_b - \
                        0.5*curr_c)
curr_beta = math.sqrt(2.0/3.0)*math.sqrt(3)*(curr_b - \
                        curr_c)/2.0

curr_d = curr_alpha*math.cos(pll_phase_angle) + \
                        curr_beta*math.sin(pll_phase_angle)
curr_q = -curr_alpha*math.sin(pll_phase_angle) + \
                        curr_beta*math.cos(pll_phase_angle)

curr_d_err = curr_ref_d - curr_d
```

```
curr_q_err = curr_ref_q - curr_q

curr_d_int += curr_d_err*dt_sample
curr_q_int += curr_q_err*dt_sample

comp_output_d = (curr_d_err*curr_kp + curr_d_int*curr_ki)
comp_output_q = (curr_q_err*curr_kp + curr_q_int*curr_ki)

comp_output_alpha = comp_output_d*math.cos(pll_phase_angle) - \
                    comp_output_q*math.sin(pll_phase_angle)
comp_output_beta = comp_output_d*math.sin(pll_phase_angle) + \
                    comp_output_q*math.cos(pll_phase_angle)

comp_output_a = math.sqrt(2.0/3.0)*comp_output_alpha
comp_output_b = math.sqrt(2.0/3.0)*(-0.5*comp_output_alpha + \
                math.sqrt(3)*comp_output_beta/2.0)
comp_output_c = math.sqrt(2.0/3.0)*(-0.5*comp_output_alpha - \
                math.sqrt(3)*comp_output_beta/2.0)
```

The references for the compensator current are provided by comp_reference.py where the VariableStorage types comp_curr_d and comp_curr_q are computed. The reference Ifd* is curr_ref_d, while Ifq* is curr_ref_q. curr_ref_q is assigned to comp_curr_q, while curr_ref_d is set to zero. This has been explained above as the compensator will supply all the reactive power demanded by the load but will not supply any active power. The three-phase currents of the compensator are curr_a, curr_b, and curr_c which will have to be measured outputs from Ammeters. These three-phase currents are transformed to the rotating d-q domain. In the rotating d-q domain, PI controllers are implemented for both d and q components. The advantage of doing so is to reduce the steady-state error in the controller. As stated before, after the PLL has synchronized to the grid, any measurements of three-phase quantities

Table 5.13 Compensator parameters

Controlled-VoltageSource	Vcompa	1C	Positive polarity toward (cell) = 1D, name of control signal = compvolta
Controlled-VoltageSource	Vcompb	9C	Positive polarity toward (cell) = 9D, name of control signal = compvoltb
Controlled-VoltageSource	Vcompc	17C	Positive polarity toward (cell) = 17D, name of control signal = compvoltc
Inductor	Lcompa	1J	0.001
Inductor	Lcompb	9J	0.001
Inductor	Lcompa	17J	0.001
Resistor	Rcompa	1H	0.1
Resistor	Rcompb	9H	0.1
Resistor	Rcompc	17H	0.1
Ammeter	Icompa	1M	Positive polarity toward (cell) = 1N
Ammeter	Icompa	9M	Positive polarity toward (cell) = 9N
Ammeter	Icompa	17M	Positive polarity toward (cell) = 17N

Table 5.14 Compensator parameters

Switch	Disconnect1a	1R	Voltage level (V) = 600.0, negative polarity toward (cell) = 1Q, name of control signal = Disconnect1a
Switch	Disconnect2a	4R	Voltage level (V) = 600.0, negative polarity toward (cell) = 4S, name of control signal = Disconnect2a
Switch	Disconnect1b	9R	Voltage level (V) = 600.0, negative polarity toward (cell) = 9Q, name of control signal = Disconnect1b
Switch	Disconnect2b	12R	Voltage level (V) = 600.0, negative polarity toward (cell) = 12S, name of control signal = Disconnect2b
Switch	Disconnect1c	17R	Voltage level (V) = 600.0, negative polarity toward (cell) = 17Q, name of control signal = Disconnect1c
Switch	Disconnect2c	20R	Voltage level (V) = 600.0, negative polarity toward (cell) = 20S, name of control signal = Disconnect2c

when transformed to the d-q reference frame will be constant d and q components as long as the three-phase signals are balanced and sinusoidal. Since this circuit is balanced and linear, any three-phase signal measured, whether they be voltages or currents, upon transformation will become dc signals in the d-q reference frame. When a PI controller is applied to dc signals, the steady-state error is ideally zero. Therefore, this implementation of PI controllers in the d-q synchronous reference frame results in zero steady-state tracking error. This is under the assumption that measurement offsets do not exist, which is true in case of a simulation as this one. The outputs of the PI controllers are the compensator voltages in the d-q reference frame. An inverse transformation is performed to transform these voltages back to the three-phase stationary reference frame so that they can be fed to the controllable voltage sources in the circuit. comp_output_a, comp_output_b, and comp_output_c have to be linked to these voltages. In this case, the output of a control function is directly impacting the circuit. Therefore, there will be three variables that will not be of the type VariableStorage but rather will be of the type Output.

The next set of variables are those at the head of the control function. The first is the time period of the control function. As before, a frequency of 10 kHz is sufficient for this controller as the fundamental frequency of the system is 60 Hz. Therefore, the time period is chosen to be 100 μs. There are also the P and I gains of the PI controllers. For this simulation, a simple trial-and-error method has been used to choose these control gains. However, vast theory exists on how these control gains can be chosen to achieve the best control performance. The user is encouraged to tweak these gains or completely redesign the control gains to verify their theory.

```
dt_sample=100.0e-6
curr_kp = 0.04
curr_ki = 8000.0

if t_clock>=t1:
        <Above code here>
        t1 += dt_sample
```

The output of the control function will be:

```
comp_outputvolt_a = comp_output_a
comp_outputvolt_b = comp_output_b
comp_outputvolt_c = comp_output_c
```

For debugging the control function and also for the purpose of examining the performance of the controller, the following variables will be declared of type VariableStorage and assigned control variables:

```
currcon_curr_d = curr_d
currcon_curr_q = curr_q
currcon_curr_d_int = curr_d_int
currcon_curr_q_int = curr_q_int
currcon_md = comp_output_d
currcon_mq = comp_output_q
currcon_ma = comp_output_a
currcon_mb = comp_output_b
currcon_mc = comp_output_c
```

There is another aspect of the control function that is not immediately apparent. There are disconnect Switches that separate the compensator from the system. It will always be necessary to disconnect the VAR compensator or for that matter any equipment from the system. In this simulation, the disconnect Switches perform another role. As can be seen from the output of the PLL, it takes around 0.15 s for the PLL to synchronize with the grid and for the angular frequency output of the PLL to achieve a steady state. Until 0.1 s, the output of the PLL is not very close to the angular frequency of the grid. Therefore, it would be advisable to wait for 0.1 s for the PLL to synchronize before performing any form of compensation. A simple timer is one option to ensure that control algorithms follow in a sequence. A more sophisticated algorithm to check if the PLL has synchronized is to check if the q component of the PLL voltage remains within a defined margin around zero as that is the indication of the PLL having synchronized. The timer code is shown below:

```
if t_clock <0.1:
        Disconnect1a = 0.0
        Disconnect2a = 0.0
        Disconnect1b = 0.0
        Disconnect2b = 0.0
        Disconnect1c = 0.0
        Disconnect2c = 0.0
else:
        Disconnect1a = 1.0
        Disconnect2a = 1.0
        Disconnect1b = 1.0
        Disconnect2b = 1.0
        Disconnect1c = 1.0
        Disconnect2c = 1.0
```

In the above block, the variables Disconnectxa, Disconnectxb, and Disconnectxc are Output variables that control the turning on and off of Switches. The reason for there being two Switches is that a Switch can only conduct in one direction—from its positive terminal to its negative terminal. Therefore, an antiparallel pair of Switches is needed in each phase to disconnect the VAR compensator while also conducting in both directions when ON.

The descriptor file for this control function currcon_sources.py is listed in Tables 5.15, 5.16, and 5.17. A brief note needs to be made about the choice of variables in control functions. As can be noticed, there are several variables common between the control functions. For example, curr_a is used in comp_reference.py and currcon_sources.py as Inputs from Ammeters. Also, curr_alpha and curr_beta are used in multiple control files. This is perfectly acceptable. The Inputs and StaticVariables are local to a control function, and multiple control functions can use variables of these types with the same name. This also makes it convenient as quite often control code is common and is merely copied from one function to another. If StaticVariables and Inputs needed to be different, the control code will need to be edited every time it is copied from one control function to another.

Outputs are a different story. When a variable is defined as an Output, it is linked to a control tag in a circuit component and therefore changes the circuit. It is possible to use the same variable for an Output within different control functions, but it is advisable that only one control function changes the control tag in the circuit component. This is for the simple reason that if multiple control functions attempt to change the state of a Switch, there could be an error in the overall control since one control function might be trying to turn it on while another might be trying to turn it off. However, the Output variable used inside a control function is local to the control function. For example, volt_a could be used in multiple control functions as long as the control functions attempt to change the voltages of different controllable voltage sources. In any case, the only variables that can be defined only once in a control function descriptor are VariableStorage types. Once a variable has been defined as a VariableStorage type in one control function descriptor, a repeat definition in another descriptor is a violation and will cause an error. This is because a variable once defined as type VariableStorage is available in all control functions—can be read and changed in any control function.

Now, we shall describe the simulation results with the VAR compensator comprised of controllable voltage sources. Figure 5.18 contains the plot of the tracking performance of the compensator with respect to the d component of current injected. Until 0.1 s, the d component of compensator current is absolutely zero as the disconnect Switches are off while the PLL synchronizes. At 0.1 s, there is a transient when the compensator connects to the grid. The steady-state d component of current supplied by the compensator is zero as the reference Ifd* = 0.

Figure 5.19 contains the plot of the tracking performance of the q component of the current of the compensator. As before, until 0.1 s, Ifq which is the green waveform is zero. After 0.1 s, when the compensator connects to the grid, after a brief transient, Ifq follows the reference Ifq* which is the red waveform and in this case is the entire q component of the load current.

Table 5.15 Compensator descriptor

Input	Element name in circuit spreadsheet = Ammeter_Icompa, desired variable name in control code = curr_a
Input	Element name in circuit spreadsheet = Ammeter_Icompb, desired variable name in control code = curr_b
Input	Element name in circuit spreadsheet = Ammeter_Icompc, desired variable name in control code = curr_c
Output	Element name in circuit spreadsheet = ControlledVoltageSource_Vcompa, control tag defined in parameters spreadsheet = compvolta, desired variable name in control code = comp_outputvolt_a, initial output value = 0.0
Output	Element name in circuit spreadsheet = ControlledVoltageSource_Vcompb, control tag defined in parameters spreadsheet = compvoltb, desired variable name in control code = comp_outputvolt_b, initial output value = 0.0
Output	Element name in circuit spreadsheet = ControlledVoltageSource_Vcompc, control tag defined in parameters spreadsheet = compvoltc, desired variable name in control code = comp_outputvolt_c, initial output value = 0.0
Output	Element name in circuit spreadsheet = Switch_Disconnect1a, control tag defined in parameters spreadsheet = Disconnect1a, desired variable name in control code = Disconnect1a, initial output value = 0.0
Output	Element name in circuit spreadsheet = Switch_Disconnect2a, variable name in control code = Disconnect2a, initial output value = 0.0
Output	Element name in circuit spreadsheet = Switch_Disconnect1b, control tag defined in parameters spreadsheet = Disconnect1b, desired variable name in control code = Disconnect1b, initial output value = 0.0
Output	Element name in circuit spreadsheet = Switch_Disconnect2b, control tag defined in parameters spreadsheet = Disconnect2b, desired variable name in control code = Disconnect2b, initial output value = 0.0

The plot in Fig. 5.20 below shows the effect of the currents injected by the compensator on the system. Plotted are the phase a source current, phase a compensator current, and phase a load voltage scaled down by a factor of 5. Let us examine the currents supplied by the three-phase source before and after the compensator is connected. The source current lags behind the phase a grid voltage as the source supplies both the active and reactive power demanded by the load when the compensator is disconnected as shown in Fig. 5.21.

As shown in Fig. 5.22, after the compensator is connected, the reactive power demanded from the source decreases and becomes zero as the compensator now supplies the load reactive power demand. As a result, the source current can be seen to have a decreased magnitude and is in phase with the load voltage while the compensator current is in quadrature with the load voltage. The simulation has allowed us to tune the control gain parameters to achieve rapid performance. These control gains can now be applied to the final simulation with the VSC where the addition is only the pulse width modulator for the VSC. It should be noted that the control gains for this final stage will be scaled down as the output of the current controller will be a modulation signal rather than the actual compensator voltages.

Table 5.16 Compensator descriptor

TimeEvent	Desired variable name in control code = t1, first time event = 0.0
StaticVariable	Desired variable name in control code = curr_d_int, initial value of variable = 0.0
StaticVariable	Desired variable name in control code = curr_q_int, initial value of variable = 0.0
StaticVariable	Desired variable name in control code = comp_output_alpha, initial value of variable = 0.0
StaticVariable	Desired variable name in control code = comp_output_beta, initial value of variable = 0.0
StaticVariable	Desired variable name in control code = comp_output_a, initial value of variable = 0.0
StaticVariable	Desired variable name in control code = comp_output_b, initial value of variable = 0.0
StaticVariable	Desired variable name in control code = comp_output_c, initial value of variable = 0.0
StaticVariable	Desired variable name in control code = comp_output_d, initial value of variable = 0.0
StaticVariable	Desired variable name in control code = comp_output_q, initial value of variable = 0.0
StaticVariable	Desired variable name in control code = curr_ref_d, initial value of variable = 0.0
StaticVariable	Desired variable name in control code = curr_ref_q, initial value of variable = 0.0
StaticVariable	Desired variable name in control code = curr_d, initial value of variable = 0.0
StaticVariable	Desired variable name in control code = curr_q, initial value of variable = 0.0

5.6 Final Stage of Control Development—The Entire Circuit with the VSC

With the current controller designed using controllable voltage sources, we reach the final stage of the simulation—simulation of the VAR compensator with the VSC. The circuit_inputs.csv file now contains additional circuit schematics and control functions as listed in Table 5.18. The change in this circuit is that the VSC is in a spreadsheet while the filter is in another separate spreadsheet. In the previous case, the controllable voltage sources along with the filter were in the same spreadsheet comp_sources.csv. The overall structure of the compensator has not changed (Tables 5.19 and 5.20).

The VSC in the spreadsheet comp_inverter.csv has parameters in comp_inverter_par-ams.csv listed in Tables 5.4 and 5.5. Each Switch in the VSC has a unique control signal name and this has been made as verbose as possible to ensure that as the circuit expands, there are lower chances of two Switches having the same control signal tag. In this circuit spreadsheet, the most important check is to ensure that the Diodes

Table 5.17 Compensator descriptor

Output	Element name in circuit spreadsheet = Switch_Disconnect1c, control tag defined in parameters spreadsheet = Disconnect1c, desired variable name in control code = Disconnect1c, initial output value = 0.0
Output	Element name in circuit spreadsheet = Switch_Disconnect2c, control tag defined in parameters spreadsheet = Disconnect2c, desired variable name in control code = Disconnect2c, initial output value = 0.0
VariableStorage	Desired variable name in control code = currcon_curr_d, initial value of variable = 0.0, plot variable in output file = yes
VariableStorage	Desired variable name in control code = currcon_curr_q, initial value of variable = 0.0, plot variable in output file = yes
VariableStorage	Desired variable name in control code = currcon_md, initial value of variable = 0.0, plot variable in output file = yes
VariableStorage	Desired variable name in control code = currcon_mq, initial value of variable = 0.0, plot variable in output file = yes
VariableStorage	Desired variable name in control code = currcon_ma, initial value of variable = 0.0, plot variable in output file = yes
VariableStorage	Desired variable name in control code = currcon_mb, initial value of variable = 0.0, plot variable in output file = yes
VariableStorage	Desired variable name in control code = currcon_mc, initial value of variable = 0.0, plot variable in output file = yes

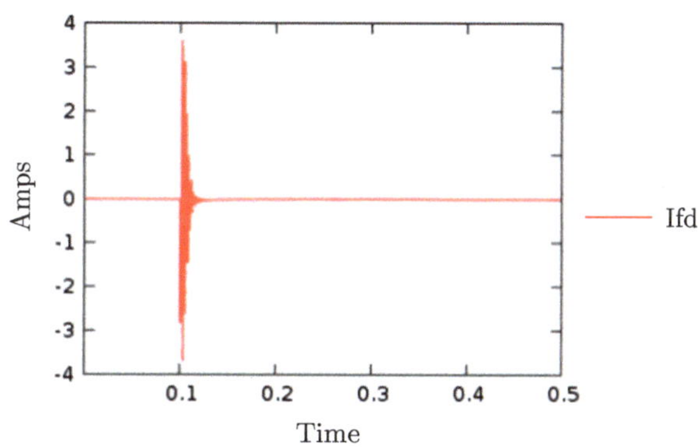

Fig. 5.18 d component of compensator current

are antiparallel to the Switches. The parameters of the filter in comp_filter.csv are in comp_filter_params.csv and are listed in Table 5.21. The control logic is quite similar and can be shown in the block diagram of Fig. 5.23. The main difference is that the outputs of the controller are the modulation signals ma, mb, and mc instead of the voltages of the controllable voltage sources. These modulation signals are signals that have a maximum magnitude of unity. Therefore, the PI control gains have to be scaled down. The control function for this current controller is currcon_inverter.py. The modulation signals are used to generate switching signals for the switches S1–S6. The modulation signal ma when compared with a triangular waveform results

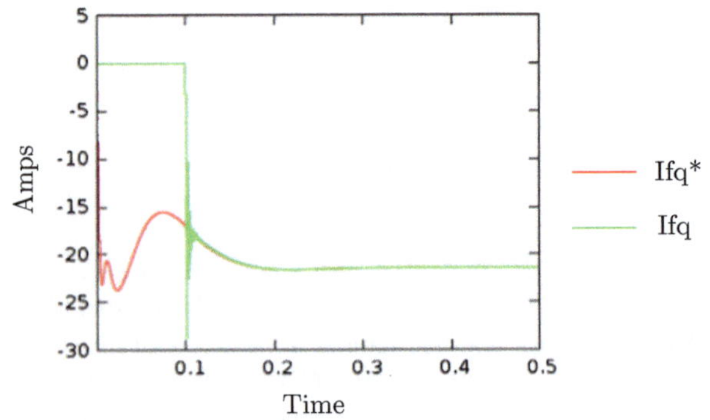

Fig. 5.19 Tracking performance of the q component of compensator current

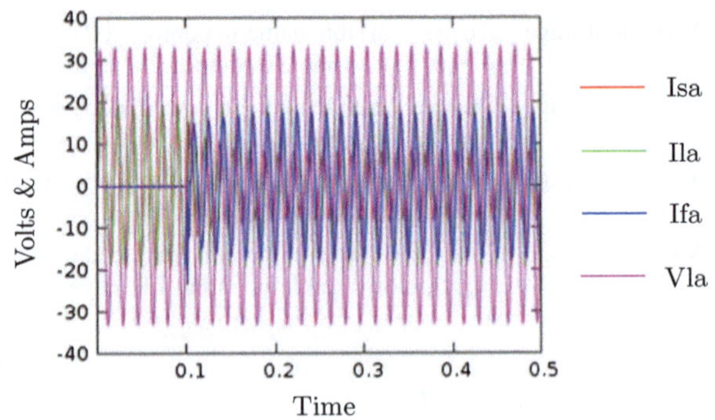

Fig. 5.20 Load voltage, source current, and compensator current

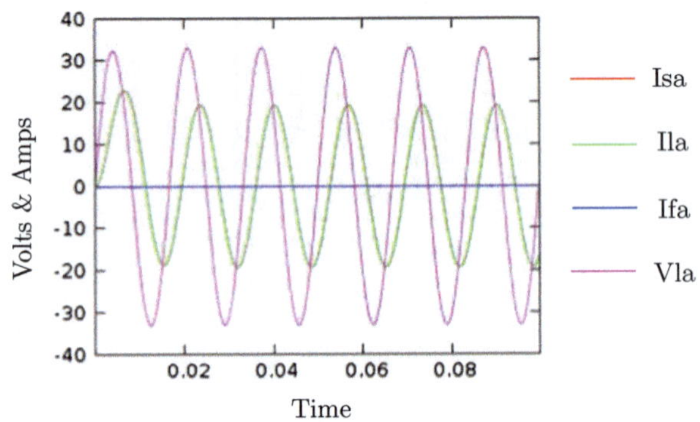

Fig. 5.21 Load voltage, source current, and compensator current

in the switching signals for S1 and S2. This is shown in the diagram of Fig. 5.24 [5]. The triangular waveform is the carrier signal. It has the frequency equal to the desired switching frequency of the VSC. It is restricted between -1 and 1. In a similar manner, modulation signal mb compared with the same carrier signal generates the switching signals for S3 and S4, and modulation signal mc compared with this carrier signal controls switches S5 and S6. The control function for the modulator is in modulator.py.

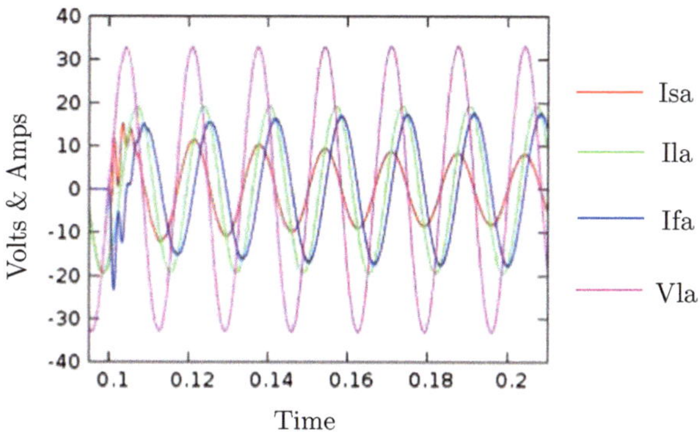

Fig. 5.22 Load voltage, source current, and compensator current

Table 5.18 Simulation parameters

Name of the circuit file	three_ph_source.csv, three_ph_load.csv, comp_i-nverter.csv, comp_filter.csv
Time duration of simulation	0.5
Time step of the simulation	1.0e-6
Time step of data storage	10.0e-6
Name of the data storage file	ckt_output.dat
Name of control files	pll.py, comp_reference.py, currcon_inverter.py, modulator.py
Split the output file?	Yes
Length of time windows	0.25

Table 5.19 VSC parameters

VoltageSource	invdcbus	15A	Peak (Volts) = 0.0, frequency (Hertz) = 0.0, phase (degrees) = 0.0, dc offset = 400.0, positive polarity toward (cell) = 14A
Resistor	invdcbus	8A	0.1
Switch	S1compinv	8H	Voltage level (V) = 600.0, negative polarity toward (cell) = 9H, name of control signal = S1compinvgate
Switch	S2compinv	22H	Voltage level (V) = 600.0, negative polarity toward (cell) = 23H, name of control signal = S2compinvgate
Switch	S3compinv	8O	Voltage level (V) = 600.0, negative polarity toward (cell) = 9O, name of control signal = S3compinvgate
Switch	S4compinv	22O	Voltage level (V) = 600.0, negative polarity toward (cell) = 23O, name of control signal = S4compinvgate
Switch	S5compinv	8T	Voltage level (V) = 600.0, negative polarity toward (cell) = 8T, name of control signal = S5compinvgate
Switch	S6compinv	22T	Voltage level (V) = 600.0, negative polarity toward (cell) = 23T, name of control signal = S6compinvgate

Table 5.20 VSC parameters

Diode	D1compinv	8J	Voltage level (V) = 600.0, cathode polarity toward (cell) = 7J
Diode	D2compinv	22J	Voltage level (V) = 600.0, cathode polarity toward (cell) = 21J
Diode	D3compinv	8Q	Voltage level (V) = 600.0, cathode polarity toward (cell) = 7Q
Diode	D4compinv	22Q	Voltage level (V) = 600.0, cathode polarity toward (cell) = 21Q
Diode	D5compinv	8V	Voltage level (V) = 600.0, cathode polarity toward (cell) = 7V
Diode	D6compinv	22V	Voltage level (V) = 600.0, cathode polarity toward (cell) = 21V

Table 5.21 Filter parameters

Inductor	Lcompa	1J	0.001
Inductor	Lcompb	9J	0.001
Inductor	Lcompc	17J	0.001
Resistor	Rcompa	1H	0.1
Resistor	Rcompb	9H	0.1
Resistor	Rcompc	17H	0.1
Ammeter	Icompa	1M	Positive polarity toward (cell) = 1N
Ammeter	Icompb	9M	Positive polarity toward (cell) = 9N
Ammeter	Icompc	17M	Positive polarity toward (cell) = 17N
Switch	Disconne-ct1a	1R	Voltage level (V) = 600.0, negative polarity toward (cell) = 1Q, name of control signal = Disconnect1a
Switch	Disconne-ct2a	4R	Voltage level (V) = 600.0, negative polarity toward (cell) = 4S, name of control signal = Disconnect2a
Switch	Disconne-ct1b	9R	Voltage level (V) = 600.0, negative polarity toward (cell) = 9Q, name of control signal = Disconnect1b
Switch	Disconne-ct2b	12R	Voltage level (V) = 600.0, negative polarity toward (cell) = 12S, name of control signal = Disconnect2b
Switch	Disconne-ct1c	17R	Voltage level (V) = 600.0, negative polarity toward (cell) = 17Q, name of control signal = Disconnect1c
Switch	Disconne-ct2c	20R	Voltage level (V) = 600.0, negative polarity toward (cell) = 20S, name of control signal = Disconnect2c

The current controller code remains the same in currcon_inverter.py with respect to currcon_sources.py. The difference is in the PI control gains and the fact that the output of the PI controller does not directly change the circuit.

```
curr_kp = 0.0001
curr_ki = 40.0
if t_clock>=t1:
    curr_ref_d = 0.0
    curr_ref_q = comp_curr_q

    curr_alpha = math.sqrt(2.0/3.0)*(curr_a - 0.5*curr_b - \
```

```
                    0.5*curr_c)
    curr_beta = math.sqrt(2.0/3.0)*math.sqrt(3)*(curr_b - \
                    curr_c)/2.0
    curr_d = curr_alpha*math.cos(pll_phase_angle) + \
                    curr_beta*math.sin(pll_phase_angle)
    curr_q = -curr_alpha*math.sin(pll_phase_angle) + \
                    curr_beta*math.cos(pll_phase_angle)

    curr_d_err = curr_ref_d - curr_d
    curr_q_err = curr_ref_q - curr_q

    curr_d_int += curr_d_err*dt_sample
    curr_q_int += curr_q_err*dt_sample
    if (curr_ki*curr_d_int > 2.5):
        curr_d_int = 2.5/curr_ki
    if (curr_ki*curr_d_int < -2.5):
        curr_d_int = -2.5/curr_ki
    if (curr_ki*curr_q_int > 2.5):
        curr_q_int = 2.5/curr_ki
    if (curr_ki*curr_q_int < -2.5):
        curr_q_int = -2.5/curr_ki

    mod_d = (curr_d_err*curr_kp + curr_d_int*curr_ki)
    mod_q = (curr_q_err*curr_kp + curr_q_int*curr_ki)

    mod_alpha = mod_d*math.cos(pll_phase_angle) - \
                mod_q*math.sin(pll_phase_angle)
    mod_beta = mod_d*math.sin(pll_phase_angle) + \
                mod_q*math.cos(pll_phase_angle)

    modsignal_a = math.sqrt(2.0/3.0)*mod_alpha
    modsignal_b = math.sqrt(2.0/3.0)*(-0.5*mod_alpha + \
                    math.sqrt(3)*mod_beta/2.0)
    modsignal_c = math.sqrt(2.0/3.0)*(-0.5*mod_alpha - \
                    math.sqrt(3)*mod_beta/2.0)

    if modsignal_a >0.97:
        modsignal_a = 0.97
    if modsignal_a <-0.97:
        modsignal_a = -0.97
    if modsignal_b >0.97:
        modsignal_b = 0.97
    if modsignal_b <-0.97:
        modsignal_b = -0.97
    if modsignal_c >0.97:
        modsignal_c = 0.97
    if modsignal_c <-0.97:
        modsignal_c = -0.97

    t1=t1+dt_sample

if t_clock <0.1:
    Disconnect1a = 0.0
    Disconnect2a = 0.0
    Disconnect1b = 0.0
    Disconnect2b = 0.0
    Disconnect1c = 0.0
    Disconnect2c = 0.0
else:
```

```
    Disconnect1a = 1.0
    Disconnect2a = 1.0
    Disconnect1b = 1.0
    Disconnect2b = 1.0
    Disconnect1c = 1.0
    Disconnect2c = 1.0

currcon_curr_d = curr_d
currcon_curr_q = curr_q
currcon_curr_d_int = curr_d_int
currcon_curr_q_int = curr_q_int
currcon_md = mod_d
currcon_mq = mod_q
currcon_ma = modsignal_a
currcon_mb = modsignal_b
currcon_mc = modsignal_c
```

As can be seen from the above code, the currcon_inverter.py function also controls the disconnect switches which in this simulation case is in the file comp_filter.csv. It is possible to create a special control function only for the disconnect switches to completely separate the reconnection and disconnection logic from the current controller. The descriptor currcon_inverter_desc.csv is a minor modification of the currcon_sources_desc.csv and is listed in Tables 5.22, 5.23, and 5.24. As can be seen, the outputs to the controllable voltage sources are eliminated. The modulation signals are stored in currcon_ma, currcon_mb, and currcon_mc as variables of the VariableStorage type. In this manner, they are made available to the modulation function described next.

The control code for the modulator is however different in several aspects. The major difference is in the sampling time of the modulator function. To fully understand this, let us examine the time periods of the signals that the modulator deals with. The modulator accepts as inputs three modulation signals ma, mb, and mc. In this

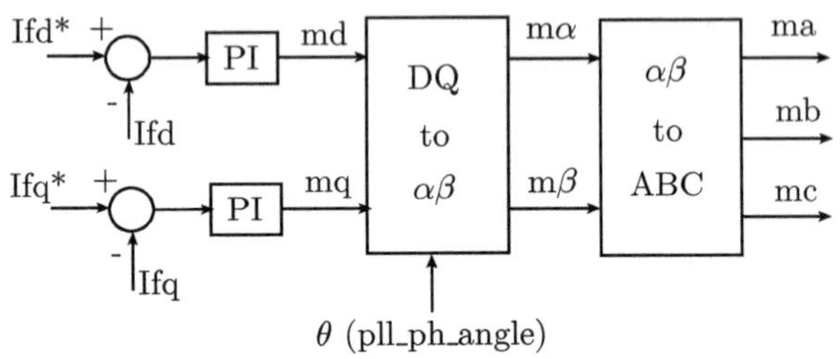

Fig. 5.23 Current control for a VSC

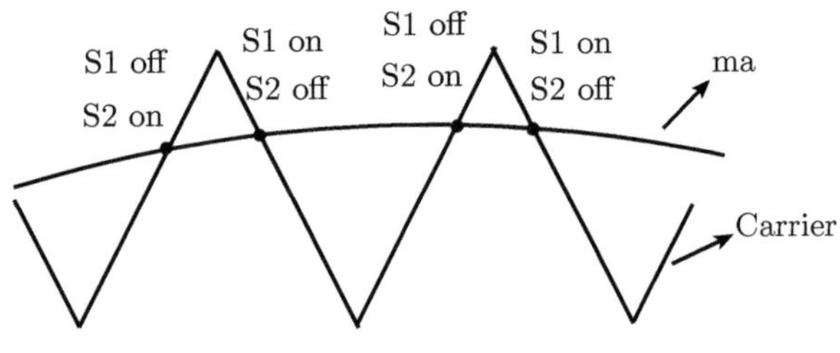

Fig. 5.24 Pulse width modulation

Table 5.22 Inverter current controller

Input	Element name in circuit spreadsheet = Ammeter_Icompa, desired variable name in control code = curr_a
Input	Element name in circuit spreadsheet = Ammeter_Icompb, desired variable name in control code = curr_b
Input	Element name in circuit spreadsheet = Ammeter_Icompc, desired variable name in control code = curr_c
Output	Element name in circuit spreadsheet = Switch_Disconnect1a, control tag defined in parameters spreadsheet = Disconnect1a, desired variable name in control code = Disconnect1a, initial output value = 0.0
Output	Element name in circuit spreadsheet = Switch_Disconnect2a, control tag defined in parameters spreadsheet = Disconnect2a, desired variable name in control code = Disconnect2a, initial output value = 0.0
Output	Element name in circuit spreadsheet = Switch_Disconnect1b, control tag defined in parameters spreadsheet = Disconnect1b, desired variable name in control code = Disconnect1b, initial output value = 0.0
Output	Element name in circuit spreadsheet = Switch_Disconnect2b, control tag defined in parameters spreadsheet = Disconnect2b, desired variable name in control code = Disconnect2b, initial output value = 0.0

Table 5.23 Inverter current controller

Output	Element name in circuit spreadsheet = Switch_Disconnect1c, control tag defined in parameters spreadsheet = Disconnect1c, desired variable name in control code = Disconnect1c, initial output value = 0.0
Output	Element name in circuit spreadsheet = Switch_Disconnect2c, control tag defined in parameters spreadsheet = Disconnect2c, desired variable name in control code = Disconnect2c, initial output value = 0.0
StaticVariable	Desired variable name in control code = curr_d_int, initial value of variable = 0.0
StaticVariable	Desired variable name in control code = curr_q_int, initial value of variable = 0.0
StaticVariable	Desired variable name in control code = mod_signal_alpha, initial value of variable = 0.0
StaticVariable	Desired variable name in control code = mod_signal_beta, initial value of variable = 0.0
StaticVariable	Desired variable name in control code = mod_signal_a, initial value of variable = 0.0
StaticVariable	Desired variable name in control code = mod_signal_b, initial value of variable = 0.0
StaticVariable	Desired variable name in control code = mod_signal_c, initial value of variable = 0.0
StaticVariable	Desired variable name in control code = mod_signal_d, initial value of variable = 0.0
StaticVariable	Desired variable name in control code = mod_signal_q, initial value of variable = 0.0

Table 5.24 Inverter current controller

StaticVariable	Desired variable name in control code = curr_ref_d, initial value of variable = 0.0
StaticVariable	Desired variable name in control code = curr_ref_q, initial value of variable = 0.0
StaticVariable	Desired variable name in control code = curr_d, initial value of variable = 0.0
StaticVariable	Desired variable name in control code = curr_q, initial value of variable = 0.0
TimeEvent	Desired variable name in control code = t1, First time event = 0.0
VariableStorage	Desired variable name in control code = currcon_curr_d, initial value of variable = 0.0, plot variable in output file = yes
VariableStorage	Desired variable name in control code = currcon_curr_q, initial value of variable = 0.0, plot variable in output file = yes
VariableStorage	Desired variable name in control code = currcon_md, initial value of variable = 0.0, plot variable in output file = yes
VariableStorage	Desired variable name in control code = currcon_mq, initial value of variable = 0.0, plot variable in output file = yes
VariableStorage	Desired variable name in control code = currcon_ma, initial value of variable = 0.0, plot variable in output file = yes
VariableStorage	Desired variable name in control code = currcon_mb, initial value of variable = 0.0, plot variable in output file = yes
VariableStorage	Desired variable name in control code = currcon_mc, initial value of variable = 0.0, plot variable in output file = yes

simulation study, these modulation signals are of fundamental 60 Hz frequency as the compensator only supplies steady reactive power. The modulator compares these modulation signals with a triangular carrier waveform of the same frequency as the switching frequency of the VSC. In this simulation study, the switching frequency of the VSC has been taken as 5 kHz. Therefore, the time period of the triangular waveform is 200 μs. The simulation time step is 1 μs. In many cases, if the modulator is run at a time period of 1 μs, the accuracy of the comparison should be sufficiently high. However, let us examine the extreme case, when the modulation signal has a magnitude of close to unity. When this happens, the comparison of the modulation signal and the triangular waveform occurs at the peak of the triangular wave. Two successive comparisons could occur in the space of a few microseconds and if the time period of the modulator is 1 μs, there could be an error in the generation of the switching signal. To avoid this, the time period of the modulator is usually taken to be smaller than the simulation time step. For a detailed explanation on how the time scheduling of events takes place, refer to Chap. 4. At this stage, it will be reiterated that the advantage of choosing an extremely small time period for the modulator is that the resolution of the comparison increases as the modulator is run at this high frequency. However, the simulator will solve the circuit only if there is a change in the output of the modulator. This implies the modulator has changed the state of one or more switches based on the comparison of the modulation signals and the carrier waveform. This is what is desired. The control code for the modulator is below:

```
import math

carr_freq = 5000.0
dt_carr = 5.0e-7

if t_clock>=tcarr:
        if (x_tri>=1.0):
                x_tri_sign=-1.0
        if (x_tri<=-1.0):
                x_tri_sign=1.0
        x_tri+=x_tri_sign*(4.0*carr_freq)*dt_carr

        if (x_tri<currcon_ma):
                sc1logic=1.0
                sc2logic=0.0
        else:
                sc1logic=0.0
                sc2logic=1.0

        if (x_tri<currcon_mb):
                sc3logic=1.0
                sc4logic=0.0
        else:
                sc3logic=0.0
                sc4logic=1.0
        if (x_tri<currcon_mc):
                sc5logic=1.0
                sc6logic=0.0
        else:
                sc5logic=0.0
                sc6logic=1.0

        tcarr += dt_carr

sc1gate = sc1logic
sc2gate = sc2logic
sc3gate = sc3logic
sc4gate = sc4logic
sc5gate = sc5logic
sc6gate = sc6logic
modulator_compinvgate1 = sc1logic
modulator_compinvgate2 = sc2logic
modulator_compinvgate3 = sc3logic
modulator_compinvgate4 = sc4logic
modulator_compinvgate5 = sc5logic
modulator_compinvgate6 = sc6logic
modulator_carrier = x_tri
```

The above control code is run at 500 ns. This number can be changed to any value. As an example, if the user plans to implement the controller for the VSC on a microcontroller with an in-built pulse width modulator, the user can set the time period to be equal to the resolution of the microcontroller hardware. The control consists of two parts—the first is the generation of the triangular carrier waveform, and the second

Table 5.25 Modulator descriptor

Output	Element name in circuit spreadsheet = Switch_S1compinv, control tag defined in parameters spreadsheet = S1compinvgate, desired variable name in control code = sc1gate, initial output value = 0.0
Output	Element name in circuit spreadsheet = Switch_S2compinv, control tag defined in parameters spreadsheet = S2compinvgate, desired variable name in control code = sc2gate, initial output value = 0.0
Output	Element name in circuit spreadsheet = Switch_S3compinv, control tag defined in parameters spreadsheet = S3compinvgate, desired variable name in control code = sc3gate, initial output value = 0.0
Output	Element name in circuit spreadsheet = Switch_S4compinv, control tag defined in parameters spreadsheet = S4compinvgate, desired variable name in control code = sc4gate, initial output value = 0.0
Output	Element name in circuit spreadsheet = Switch_S5compinv, control tag defined in parameters spreadsheet = S5compinvgate, desired variable name in control code = sc5gate, initial output value = 0.0
Output	Element name in circuit spreadsheet = Switch_S6compinv, control tag defined in parameters spreadsheet = S6compinvgate, desired variable name in control code = sc6gate, initial output value = 0.0

Table 5.26 Modulator descriptor

StaticVariable	Desired variable name in control code = sc1logic, initial value of variable = 0.0
StaticVariable	Desired variable name in control code = sc2logic, initial value of variable = 0.0
StaticVariable	Desired variable name in control code = sc3logic, initial value of variable = 0.0
StaticVariable	Desired variable name in control code = sc4logic, initial value of variable = 0.0
StaticVariable	Desired variable name in control code = sc5logic, initial value of variable = 0.0
StaticVariable	Desired variable name in control code = sc6logic, initial value of variable = 0.0
StaticVariable	Desired variable name in control code = x_tri, initial value of variable = 0.0
StaticVariable	Desired variable name in control code = x_tri_sign, initial value of variable = 0.0
TimeEvent	Desired variable name in control code = t_carr, first time event = 0.0
VariableStorage	Desired variable name in control code = modulator_compinvgate1, initial value of variable = 0.0, plot variable in output file = yes

is the comparison of the modulation signals and the generated carrier waveform. The output of modulator.py are the switching control signals of the switches. The descriptor modulator_desc.csv is as listed in Tables 5.25, 5.26 and 5.27.

Table 5.27 Modulator descriptor

VariableStorage	Desired variable name in control code = modulator_compinvgate1, initial value of variable = 0.0, plot variable in output file = yes
VariableStorage	Desired variable name in control code = modulator_compinvgate2, initial value of variable = 0.0, plot variable in output file = yes
VariableStorage	Desired variable name in control code = modulator_compinvgate3, initial value of variable = 0.0, plot variable in output file = yes
VariableStorage	Desired variable name in control code = modulator_compinvgate4, initial value of variable = 0.0, plot variable in output file = yes
VariableStorage	Desired variable name in control code = modulator_compinvgate5, initial value of variable = 0.0, plot variable in output file = yes
VariableStorage	Desired variable name in control code = modulator_compinvgate6, initial value of variable = 0.0, plot variable in output file = yes
VariableStorage	Desired variable name in control code = modulator_carrier, initial value of variable = 0.0, plot variable in output file = yes

With a description of the control code, we will now examine the simulation results. The data written to the output files are in the following order:

```
*********************************************
Output file is in ckt_output1.dat and ckt_output2.dat and
the columns are in the following sequence
1 => Time
Meters are in the following sequence:
2  =>  Ammeter_Isource1a at  1L
3  =>  Ammeter_Isource1b at  11L
4  =>  Ammeter_Isource1c at  20L
5  =>  Ammeter_Iload1a at  1M
6  =>  Ammeter_Iload1b at  9M
7  =>  Ammeter_Iload1c at  17M
8  =>  Ammeter_Icompa at  1M
9  =>  Ammeter_Icompb at  9M
10  =>  Ammeter_Icompc at  17M
11  =>  Voltmeter_Vload1a at  36G
12  =>  Voltmeter_Vload1b at  36J
13  =>  Voltmeter_Vload1c at  36M

Control variables to be plotted are in the following sequence
14  =>  pll_phase_angle
15  =>  currcon_curr_q
16  =>  pll_cont_int
17  =>  currcon_disconnect1b
18  =>  currcon_disconnect1c
19  =>  currcon_disconnect1a
20  =>  currcon_curr_d
21  =>  pll_volt_d
22  =>  modulator_carrier
23  =>  modulator_compinvgate6
24  =>  currcon_curr_q_int
```

```
25   =>   pll_omega
26   =>   pll_volt_q
27   =>   currcon_curr_d_int
28   =>   modulator_compinvgate2
29   =>   currcon_md
30   =>   currcon_disconnect2c
31   =>   currcon_disconnect2b
32   =>   currcon_disconnect2a
33   =>   modulator_compinvgate4
34   =>   currcon_mb
35   =>   currcon_mc
36   =>   currcon_ma
37   =>   modulator_compinvgate1
38   =>   modulator_compinvgate3
39   =>   comp_curr_q
40   =>   modulator_compinvgate5
41   =>   comp_curr_d
42   =>   currcon_mq
*************************************************
```

Also it should be noted that in this case, the output is split into files having a 0.25 s time window. Since the total time duration of the simulation is 0.5 s, this would result in two output data files—ckt_output1.dat and ckt_output2.dat. This splitting

Fig. 5.25 Load voltage, source current, and compensator current

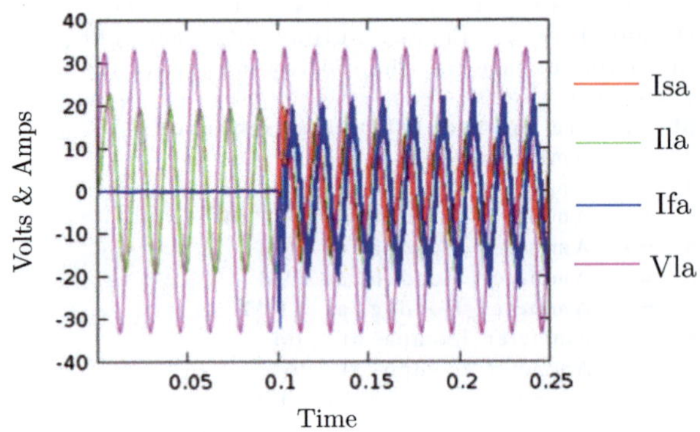

Fig. 5.26 Load voltage, source current, and compensator current

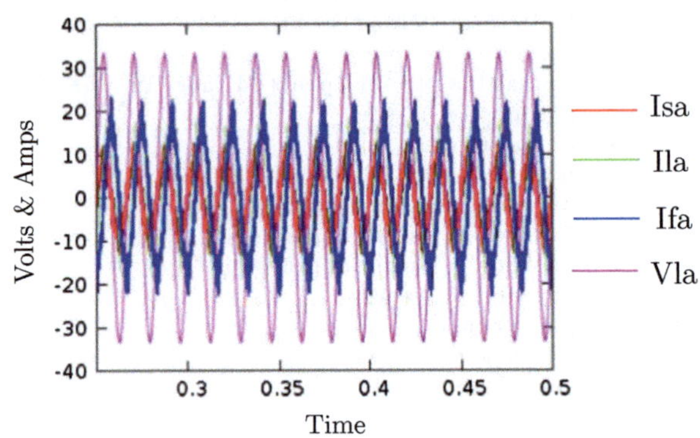

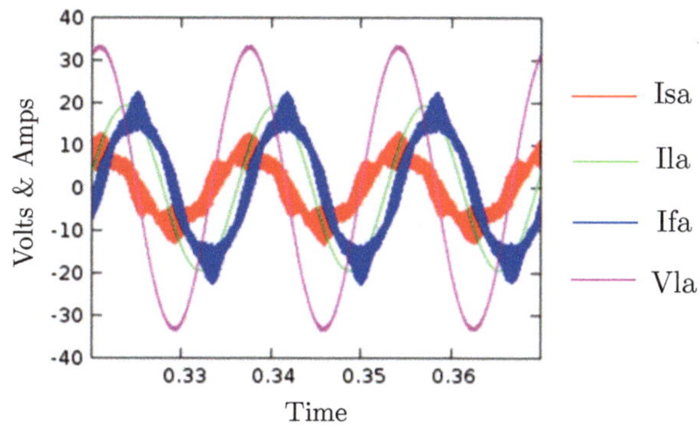

Fig. 5.27 Load voltage, source current, and compensator current—close-up

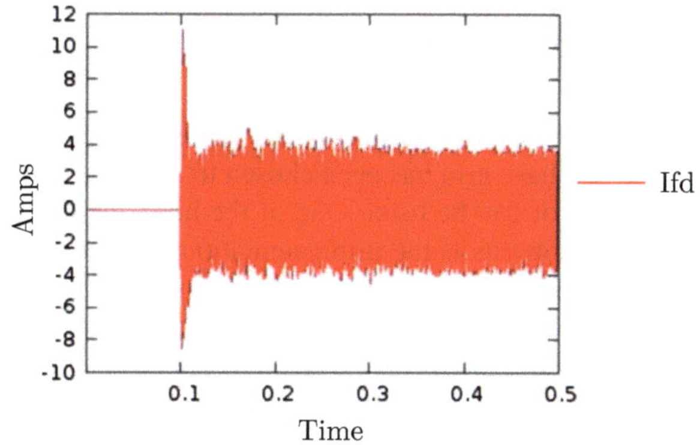

Fig. 5.28 Tracking performance of d component of compensator current

of files is not completely necessary in this case, however if running for a longer time duration, it would be advisable to split the output data files to make plotting easier and faster.

As before, we plot the phase a load voltage, source current, load current, and compensator current. For the first 0.25 s, the plot is in Fig. 5.25 while the plot for the latter 0.25 s is in Fig. 5.26. A close-up showing the performance of the compensator is shown in Fig. 5.27. The plot in Fig. 5.28 shows the d component of the current injected by the VAR compensator into the grid. As can be seen, both data items from both output files can be plotted in one figure. The user should check how much of a burden it is for the plotting software used to plot data items from multiple files. In case the software hangs, it is advisable to repeat the plotting command with fewer data items. Figure 5.29 shows the plot for the q component of the current injected by the VAR compensator. The q component of the current injected by the compensator tracks the reference which is the q component of the load current. The difference in all these plots is the switching ripple component in the currents which does not exist when the compensator has been implemented as controllable voltage sources.

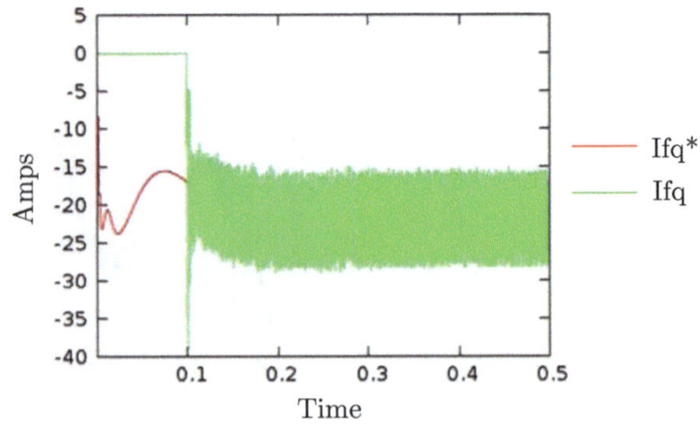

Fig. 5.29 Tracking performance of q component of compensator current

5.7 Conclusions

As stated in the introduction of this chapter, the case of a VAR compensator connected to a three-phase grid has been chosen to describe to a user how the functionalities of the simulator can be used. One of the most important aspects of simulating power electronic circuits is the implementation of controllers. It is essential that a user be able to build a simulation model as close as possible to the final hardware implementation. There are several aspects of hardware implementation that are critical to the functioning of a power converter. The first and foremost being the frequency of execution of the control function. In most control designs, the frequency of execution will be a major design constraint and will affect the final control gains. In a digital implementation, the frequency of execution is maintained by using interrupt service routines and timers or by using external oscillators. The simulator allows every control function to have multiple time events. By defining a time event for a control function, the simulator will ensure that the control function is executed at that time instant. Furthermore, by introducing the concept of an event-driven control function, a control function can be evaluated at any arbitrary instant. However, the simulator will run the circuit solvers only when a control function generates an event by changing one or more of its outputs. In this manner, a user can generate random time events in a control function that may be larger or smaller than the simulation time step. Therefore, the time event variable available with every control function allows the user to determine the exact time instant of evaluation of a control function similar to a hardware implementation.

Another significant feature of the simulator is the ability to store large amounts of data that a user can plot. Every Ammeter and Voltmeter stores the measured current and voltage in the specified output file. Moreover, every control function can store values by using the VariableStorage data type. In large simulations, the amount of data generated is quite often a factor for simulations to hang. This simulator solves the problem by allowing a user to split the output file by time intervals. The last simulation specifies a 0.25 s time interval that results in two output data files. The advantage of doing so is that a user can delete intervals that are not of importance or alternatively compress data files as they are mere text files that can be compressed to 25% of their original size with basic compression software. In this manner, all the data from a simulation is available to a user. In many software, for large simulations, the only option available to the user is to plot the last specified time interval while all data previous to that are deleted.

This chapter has shown how a circuit can be simulated in stages. It has been described how each segment of the circuit can be placed in a separate spreadsheet and connected to the rest of the circuit using jump labels. In order to include this segment of the circuit in the simulation, the spreadsheet needs to be added in the simulation parameter spreadsheet circuit_inputs.csv. In this manner, the simulation can be developed in a manner similar to commercial software where a segment of a circuit can be embedded in a block or subsystem and this can be connected to the rest of the circuit through its ports and connectors. This feature of the simulator of embedding segments of a circuit in separate spreadsheets improves the reusability of these circuit segments in other simulations.

Along with the development of a simulation by adding and deleting circuit segments, the chapter has also described how control functions can be added and removed from a simulation. As with circuit segments, a control function can be added to a simulation by adding it to the simulation parameters file circuit_inputs.csv. The importance of the TimeEvent variable has been described above. Additionally, the StaticVariable data type allows the user to perform complex mathematical calculations. This is because the value of a StaticVariable object is stored by the simulator and can be used to perform continuous summation, integration, and also implementation of higher-order filters. The examples in this chapter have described how a PI controller can be implemented using StaticVariables.

As stated in Chap. 4, objects of the VariableStorage data type can be used to exchange data between control functions. In this chapter, this feature has been used for two purposes: the first being to store control variables in the output data file and the second being to interface control functions such as the PLL and the current controller. The VariableStorage data type is an extremely powerful object that allows a user to link control functions together. This is particularly useful when power converters need to be controlled in a coordinated manner such as interleaved switching to reduce current harmonics. The VariableStorage data type can also serve as a communication channel between control functions when in a practical implementation, power converters are controlled using remote sensing or communication particularly in microgrids and distributed systems.

The example of the VAR compensator in this chapter has shown how all the different features of the circuit simulator make it comparable to existing commercial software. The major drawback of the simulator is the lack of a graphical user interface that makes it inconvenient to use particularly if the need is to simulate a fairly simple circuit. However, for large and complex circuits, a graphical user interface does not play as much of a role. For example, in a circuit with several power converters and filters, the usual strategy is to use blocks containing each subcircuit. Furthermore, the parameters of the circuit are not directly entered into the dialog boxes of each component but are entered in script files so as to keep track of changes made to the circuit. Therefore, the advantage of using this circuit simulator would be visible when circuits get larger. As stated before, ever since the simulator has been conceived as a project, the final objective has been to able to simulate large circuits for which many existing simulators are not convenient to use. The reader is encouraged to visit the Web site of the simulator to look at some of the more complex circuits that have been simulated to get an idea of the scope and possibilities of the project.

Chapter 6
Nodes, Branches, and Loops

Abstract This chapter describes how the simulator processes the circuit schematics that the user enters in spreadsheets. The connectivity information is extracted from the circuit schematics in the form of nodes, branches, and loops. Nodes, branches, and loops are used to perform circuit analysis through loop analysis and nodal analysis which are described in the next chapters. The chapter describes through sample circuits, the algorithms used to determine the nodes, branches, and loops. The chapter introduces the concept of the LoopMap which is used for performing loop analysis in Chap. 7 and the concept of KCLBranchMap which is used for performing nodal analysis in Chap. 8.

Keywords Circuit connectivity · Nodes · Branches · Loops · Jumpers · Short nodes · Short branches · Iterative search algorithms · Loop analysis · Nodal analysis

6.1 Introduction

Chapters 3, 4, and 5 have described the exterior of the circuit simulator. Chapter 3 described how the circuit schematic can be entered in a spreadsheet and how the parameters of the components can be entered in a separate spreadsheet. Chapter 4 described how a user can define control functions and how the circuit simulator processed them. Chapter 5 provided a tutorial on how a circuit with a power converter can be simulated along with all the associated control functions. These chapters, however, did not describe how the simulator solves the circuit and generates information of currents and voltages. The rest of the book is dedicated to describing the core engine of the simulator that performs circuit analysis.

The first layer of the simulation engine processes the circuit extracted from the user-entered spreadsheets and generates connectivity information. This connectivity information is in the form of nodes, branches, and loops. For any form of circuit analysis, these are essential. The circuit simulator uses both loop analysis and nodal analysis as will be described in Chaps. 7 and 8, respectively. Nodal analysis needs the information of the nodes in the circuit and the branches that connect these nodes.

S. V. Iyer, *Simulating Nonlinear Circuits with Python Power Electronics*,
https://doi.org/10.1007/978-3-319-73984-7_6

Loop analysis requires information of the number of closed paths in the circuit, which is derived from the nodes and branches in the circuit. This chapter will describe how the simulator will generate the nodes, branches, and loops in the circuit and present them in a format that can be used by subsequent chapters that describe loop analysis and nodal analysis.

The previous chapters provided detailed code samples in order to help the user be able to write control functions while simulating circuits. However, the rest of the book that describes the core simulation engine will focus on describing the concept of circuit simulation. In this chapter, the method of determining nodes, branches, and loops will be described using sample circuits. Moreover, the method used in the simulator to determine nodes, branches, and loops will be described using diagrams and charts to make it easier to understand these concepts. Since the simulator is open source, readers are encouraged to go through the source code while reading these chapters if they would like to modify the simulation engine.

6.2 Jump Labels

Jump labels were introduced in Chap. 3 while describing circuit components that a user can enter in the schematic spreadsheet. A jump label as will be described is a discontinuity in a circuit, and therefore while processing a circuit, the simulator begins by looking for jump labels. In some circuits, a jumper wire is critical if the circuit has to be described in a two-dimensional input area like a spreadsheet. An example of such a circuit is a single-phase Diode rectifier as shown in Fig. 6.1. In general, a large and complex circuit would need to be drawn in modular blocks and the blocks connected together by connectors that can be also perceived as jump labels. For that purpose, let us begin by describing the purpose of jump labels and the manner in which they can be used. A pair of jump labels in a circuit schematic are physically connected. This enables breaking a branch to enable a connection similar to the lower branch of the above Diode rectifier. There are a few limitations to using jump labels as will be described below:

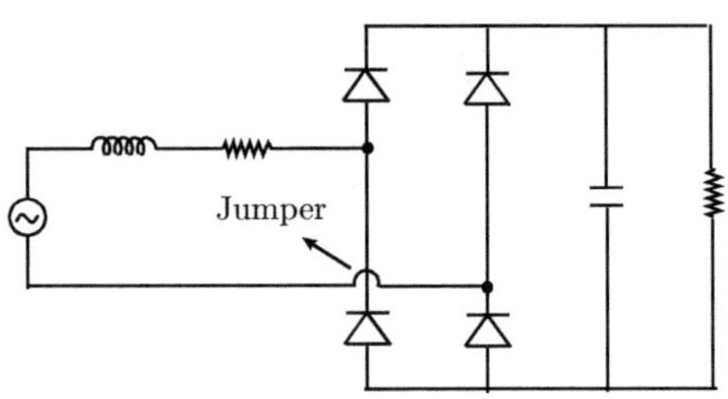

Fig. 6.1 Single-phase Diode rectifier

1. Though technically, many connections can be made in a circuit, jump labels must occur and can occur only in pairs. So, jump labels are to connect two parts of a branch segment and cannot be used to form a node with two or more branch segments. This implies that the circuit simulator will throw an error if there is only one jump label or if there are more than two of the same jump labels.
2. As jump labels are to connect two segments of a branch together, a jump label must be the extreme element in the branch segment. A jump label must have an element only on one of the cells adjacent to it as it connects this element to the other instance of the jump label. It is an error to have elements on two cells adjacent to a jump label as this would make the connection ambiguous. Similarly, a jump label cannot be next to another jump label as two jumps cannot be made immediately. At least one element has to be inserted between two jump labels. For this same reason, a jump label cannot be found next to a node as this would make the connection to a node ambiguous. At least one element has to be inserted between a node and a jump label.

The diagram in Fig. 6.2 shows the unacceptable use of jump labels on the left and their correct use on the right.

The simulator reads the entire circuit for jump labels and checks for sanity in jump labels according to the above guidelines. A jump label contains the keyword "jump" followed by the identifier. This identifier can be anything—a number, a letter, or a combination of letters and numbers. It is advisable not to use special characters. For ease in debugging the circuit at a later stage, it is advisable to let the identifiers be descriptive of the parts being connected. Therefore, "jump1", "jumpsrctodiode",

	A	B	C	D	E	F	G	H	I	J	K	L	M
1	wire	wire	wire	jump1	wire			wire	wire	wire	wire	jump1	
2													
3													
4		jump1	wire	wire	wire				jump1	wire	wire	wire	
5													
6													
7													
8	wire	wire	wire	wire	jump1	jump2				jump2			
9										wire			
10								wire	wire	wire	wire	jump1	
11		jump1	wire	wire	wire								
12				wire									
13				jump2					jump1	wire	wire	wire	
14											wire		
15											jump2		
16													
17													
18													
19	wire	wire	wire	wire	wire			wire	wire	wire	wire	wire	
20				wire							wire		
21				jump1							wire		
22				wire							jump1		
23													
24													

Fig. 6.2 Using jump labels

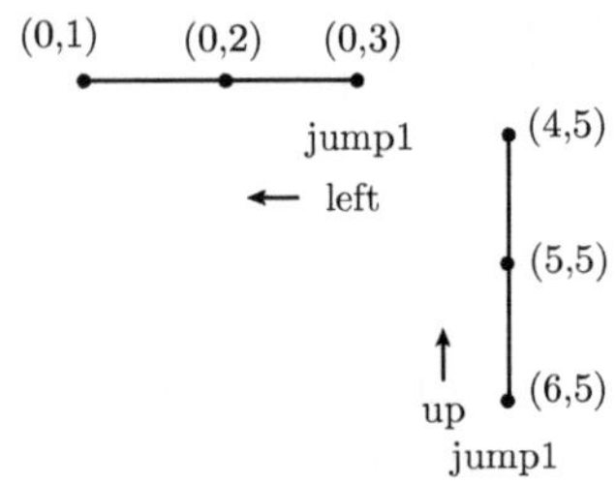

Fig. 6.3 Directions at jump labels

"jumpcaptoload" are all acceptable jump labels as long as there is another identical jump label in the circuit. All the jumps are added to a list called the JumpList which has the following structure:

```
[Row, Column, JumpLabel, Direction]
```

Row and Column form the cell position of the jump label. JumpLabel is the name of the jump label at the cell position (e.g., "jump1"). Direction is the direction in which the next cell can be found from this jump label and can be "up", "down", "right", or "left". This is to make the branch search algorithm easier upon finding a jump label. As stated before, a jump label has to be the extreme element in a branch segment. Therefore, if the only element adjacent to a jump label is one cell position below the jump label, direction is "down" indicating that when the branch search algorithm arrives at this jump label, this is the direction in which it needs to proceed to continue along the branch segment. As an example, in Fig. 6.3, the jump labels "jump1" on the branch segment on the left will be [0, 3, "jump1", "left"] as the branch segment progresses to the left when you exit at this jump label. The "jump1" label on the right will be [6, 5, "jump1", "up"] as the branch segment progresses upward when exiting at the jump label. It is to be noted again that the direction at a jump label is the direction when exiting the jump label and not the direction when entering the jump label.

The next step is to map the jump labels together in a "JumpMatrix". This "JumpMatrix" is a dictionary with keys being the jump labels, and the value associated with each key (jump label) is the following list:

```
[[[row, col], ''direction''], [[row, col], ''direction'']]
```

The list contains embedded lists—one for each jump label. Each jump label has the structure similar to the JumpList—the cell coordinates [row, col] and the exit "direction". The advantage of using a dictionary is faster access to values when the key (jump label) is known. The objective of collecting the pairs of jump labels in the above dictionary is to access the jump label when it is encountered in a branch search function and to determine the cell position from which to continue. Typically, as the circuit increases in size, search functions would become time consuming and therefore, a dictionary will reduce the time in locating jump labels as dictionary search functions are based on item keys and are optimized. The alternative would have been to continue with the previous "JumpList" but since that is a list, the method of determining jump labels during the branch search function would be to sequentially search through the list which could be time consuming if there are hundreds of jump labels. For the above example in Fig. 6.3, the "jump1" label will have the following key:value dictionary entries:

```
''jump1'': [[[0, 3], ''left''], [[6, 5], ''up'']]
```

To sum up, the dictionary "JumpMatrix" serves to map the jump labels in the circuit by making them easily accessible through unique keys which are the jump labels themselves. When a jump label is encountered, the value of the jump label is read from the dictionary and it can be determined which of the two jump labels has been encountered. Whichever jump label is encountered, the branch continues at the other jump label and furthermore, the direction associated with the jump label indicates the direction in which the branch proceeds from the jump label. This will be described in further detail in the next section related to branch determination.

6.3 Nodes and Branches

A circuit is extracted from a spreadsheet onto a matrix which in Python are lists embedded within lists. This has been described in Sect. 3.2 of Chap. 3. This matrix is called "ConnMatrix". Each element in "ConnMatrix" is iterated through. An element may be empty, or a wire or a component or a jump label. Chapter 3 focused on how objects were created for components and how the user would enter the parameters of these components. In this chapter, the simulator will process the connectivity of the circuit. Since the circuit is "drawn" on a spreadsheet, an element can have only four elements adjacent to it and these are the elements in the cells above, below, to the right, and to the left of it. A function is called that performs a check for whether the element is a node. A wire element can be a node when it is a T joint or if it is a "cross" in a circuit. A component or a jump label cannot form a node and doing so in a circuit schematic is an error. A T node is when an element has elements on two opposite directions adjacent to it and only one element on one other direction adjacent to it. For example, an element has adjacent elements above it and below it and an adjacent element to the right. Or alternatively, it has adjacent elements to the right and left of it and an adjacent element below it. A "cross" is when an element has adjacent elements on all directions—above, below, to the right, and to the left. The five types of nodes are shown in Fig. 6.4. All such nodes found in a circuit are added to a list called NodeList as [row, column] lists.

Once all the nodes are determined, the method of determining branches is to start from a node and find elements until another node is encountered. The branch then becomes the collection of elements between the two nodes and includes the two nodes. The only complication that arises in this process is the presence of jump labels and continuing along the branch when jump labels are encountered. The branches are collected together in a list called "BranchParams". Additionally, a matrix called "BranchMap" is generated which has the information of the branches that exists

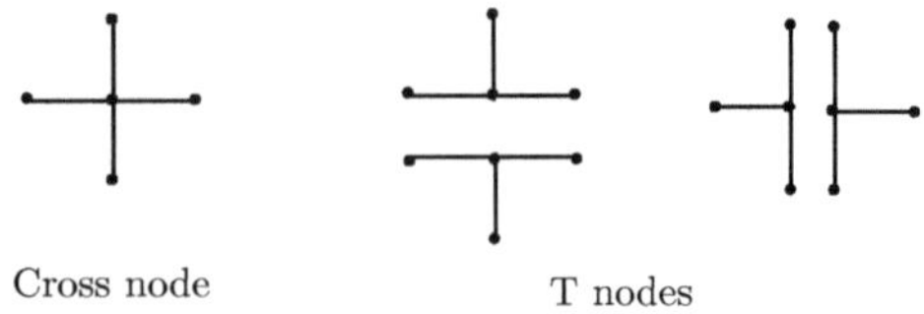

Fig. 6.4 Types of nodes

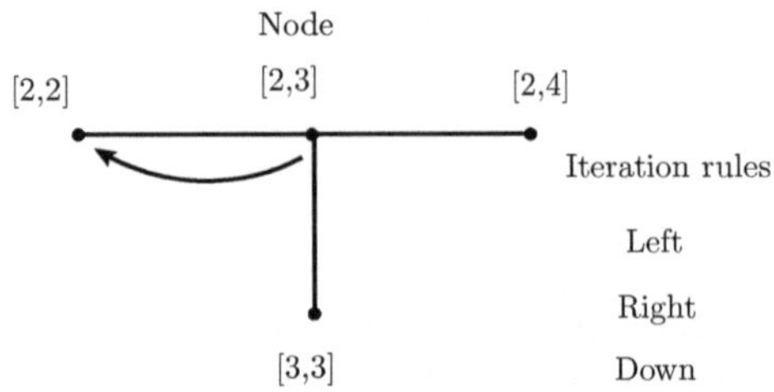

Fig. 6.5 Directions at nodes

between all the pairs of nodes. At this point, it should be noted that every branch must be completed. This means once you start a branch from a node, you must extend it all the way to another node. An incomplete or a broken branch will cause the circuit simulator to abort with an error. If the intention is to not have current flowing between two nodes, connect them with a very large resistance. Additionally, a branch is merely the existence of an element. However, it is also possible to have an empty branch, i.e., comprised only of "wire" elements. These empty branches will be dealt with later in the chapter.

Since the starting point is a node, a node by its definition has at least three elements in adjacent cells. Therefore, when starting from a node, there could potentially be at least three directions in which to search from branches and these are added to a list called "NodeIterRule". Each node has a node iteration rule that is a dictionary with four possible directions "up", "down", "right", "left". For each node, the directions in which elements exist are determined and the directions in which there are empty cells are deleted. The search algorithm iterates through all the nodes in "NodeList" and will look for branches starting from a particular node. For each iteration rule for that node, the search algorithm begins to add elements and form a branch. For example in the node [2, 3] shown in Fig. 6.5, there are three iteration rules at this node—"left", "right", and "down".

A temporary list called "BranchIter" is started for each direction in "NodeIterRule" corresponding to a node, and this list contains the elements that are added as the branch is being determined with the first element of this list being the starting node. The row and column of this first element are stored in the variables "NodeRow" and "NodeColumn", respectively. The next element will be decided based on the node search iteration rule. For example, if the current node search iteration rule is "left", this implies the next element will have the same row but the column will be one less than the column of the first element, i.e., [2, 2] (as shown by the arrow in Fig. 6.5). This row and column pair of the new element is stored in the variables NextNodeRow and NextNodeColumn, respectively. A variable to keep track of jumps is initialized to

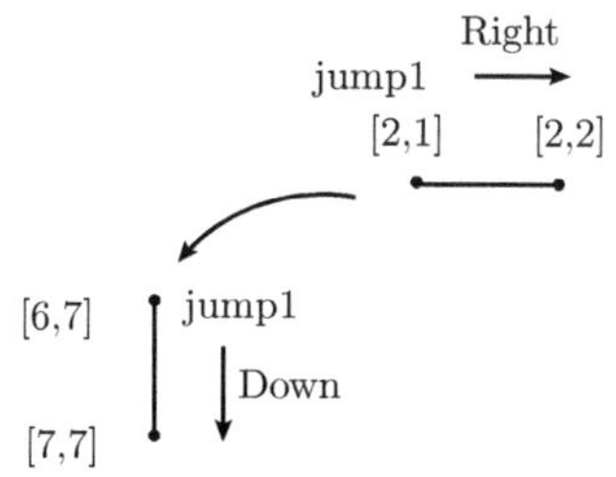

Fig. 6.6 Performing a jump during a branch search

an empty string and is called "JumpExecuted". This is in preparation for the iterative process to be started next.

The iterative process now begins, and this iterative process will end when the terminating condition is met which is that the row and column pair of [NextNodeRow, NextNodeColumn] is in the list "NodeList". This means, the next element in the branch search algorithm is a node and the branch is complete. There are two possibilities at every stage of the search algorithm—the next element [NextNodeRow, NextNodeColumn] is a jump label, or it is not a jump label. Let us consider the possibility of this next element being a jump label. Refer to Fig. 6.6 where the next element [2, 1] is the jump label "jump1". This check is performed in the function "BranchJump". Within this function, it is determined if the element is a jump label by checking if the first four characters are "jump". If the element is a jump label, the dictionary "JumpMatrix" described in the previous section is looked up. For the above example, the dictionary entry for "jump1" will be:

```
''jump1'': [[[2, 1], ''right''], [[6, 7], ''down'']]
```

As each item in the dictionary "JumpMatrix" contains two elements, it first needs to be determined which of the two elements has currently been encountered in the branch search algorithm. If the first element has been encountered, the search picks up the second element to continue the branch while if the second element has been encountered, the search picks up the first element. In the above example, the first element is encountered, and therefore, the search algorithm continues from the second element. This continuation is performed in the function "JumpMove", and the purpose of this function is to determine the element adjacent to the other jump label in the pair. Though it appears trivial, the description below ensures that branch search algorithm does not go back to the origin node. The function "JumpMove" will use the direction associated with the continuing jump label which in this case is [6, 7] and "down" to determine the next element after exiting the jump. Since the exit direction is "down", the next element after the jump will be [7, 7] as can be seen in Fig. 6.6. At this point since a jump has been executed, the "JumpExecuted" variable will be set to the value "down". Now that the jump has been executed, the continuation of the branch is performed by the function "BranchAdvance".

Since two jump labels cannot be in adjacent cells, the next element adjacent to the branch labels will not need to look for jump labels. From the example above, from element [7, 7] the search algorithm could progress in any of three directions—

left, right, or down to the elements [7, 6], [7, 8], or [8, 7], respectively. The only direction the branch search algorithm cannot progress in is “up” because that would lead the search back to the jump label “jump1” at [6, 7]. And this is precisely the need for the variable “JumpExecuted” which is currently set to “down”. Therefore, the search algorithm can look in any direction except “up”. For this purpose, when the search algorithm looks in a particular direction for a component, the check condition is whether the variable “JumpExecuted” is not set to the direction opposite to this direction. In the above example, if from [7, 7] the next element was to the left at [7, 6], the check performed is whether “JumpExecuted” is set to “right”. Since that is not the case, the next element would become [7, 6]. This check ensures a reversal does not take place in the function “BranchAdvance”. Once an adjacent element in the appropriate direction has been found, check if the element has not already been found so far, i.e., if the element is not already in the temporary list “BranchIter”. This is a repeat check to ensure that the search algorithm is not going backward to the origin node. If not, it is a new element and can be added to “BranchIter”. Now that the jump is dealt with, set the “JumpExecuted” variable to a null string “”. Since the search algorithm is at two elements away from the jump label, the current element can be another jump label which will be checked in the next iteration.

If, however, a jump label is not encountered, the function “BranchJump” will not produce any change in the next element and the “JumpExecuted” variable will remain a null string “” as it changes only when a jump is executed. This unchanged next element will be passed to the function “BranchAdvance”, and since “JumpExecuted” is a null string, this function will merely return back an element adjacent to it. Since in a branch, every element will have two adjacent elements for continuity, one of the elements will already have been encountered and will exist in “BranchIter” and therefore, the new adjacent element will be returned. This can be illustrated with Fig. 6.7. Let us assume the current element is [3, 6] and the arrows indicate the progress of the search algorithm. Since no jump has been encountered, the “JumpExecuted” variable is a null string. At [3, 6], there are two elements connected [3, 5] and [3, 7]. To progress on to [3, 5], there are two checks. Since [3, 5] is to the left of [3, 6], the first check is whether “JumpExecuted” is not “right”. Though this condition is satisfied as “JumpExecuted” is a null string, the next condition is whether the element exists in “BranchIter”. Since the element has been encountered by the search algorithm, it will exist in “BranchIter” and therefore will be rejected as a possible next element. The element [3, 7] satisfies both conditions as “JumpExecuted” is not “left” and [3, 7] is not in “BranchIter”.

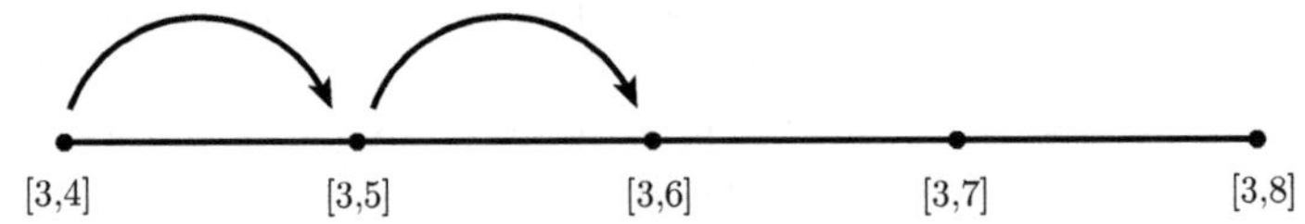

Fig. 6.7 Adding an element to a branch

In the case of a broken branch, the branch search algorithm will halt at the last element in the segment as the only element adjacent to this has been encountered before. This is an error condition and will cause the simulator to abort. If the branch is continuous and extends from one node to another node, the search algorithm will end with the terminating condition that the next element generated by the function "BranchAdvance" is a node. The temporary list "BranchIter" is now closed by adding this terminating node and adding "BranchIter" to the matrix "BranchMap" with the coordinates of "BranchMap" being the indices of the two nodes that terminate the branch. The BranchMap will be dealt with in detail when describing the loop search algorithm.

6.4 Short Branches and Nodes

While drawing a circuit, for clarity in representation, it becomes essential to have empty branches, i.e., branches without any components. For example, examine the segment of the circuit in Fig. 6.8. The above circuit could have been drawn without the empty branches indicated in the circle but using these empty branches greatly improves the clarity of the circuit. These branches contain no components and are comprised only of "wire" elements. Depending on the complexity of the circuit, these empty branches could form almost half of the total branches and therefore, leaving these branches in the list of branches and passing them off to the simulator for processing could significantly increase the burden of the simulator. The empty branches are called "short" branches, and the nodes that are connected together by "short" branches are called "short" nodes.

"Short" branches are completely eliminated from the circuit simulator but "short" nodes are retained in a separate list as they still contain the connectivity information of the circuit. For example, there are three nodes in the circuit of Fig. 6.8 that are connected by "short" branches and therefore are "short" nodes. If the "short" branches are eliminated, the connectivity information needs to be carried over as the nodes A, B, C, and D are connected together by "short" branches. Without these branches, a record needs to be maintained that these nodes are connected together and are essentially a single node. The procedure of identifying "short" branches and "short" nodes is therefore linked together.

Figure 6.9 shows a circuit that is connected together by "short" branches represented by numbers, and the resulting "short" nodes are represented by alphabets. By

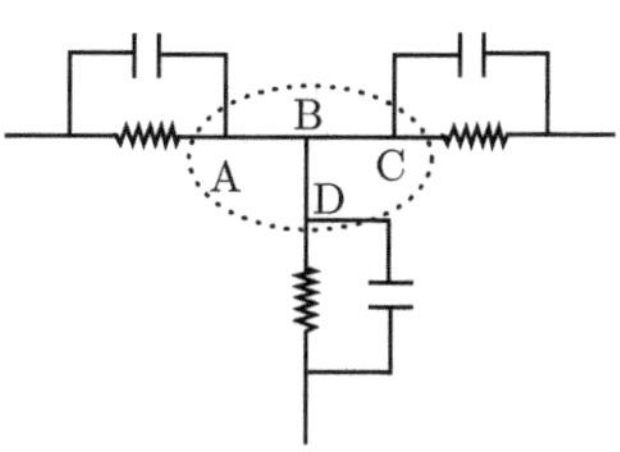

Fig. 6.8 Short nodes connected by short branches

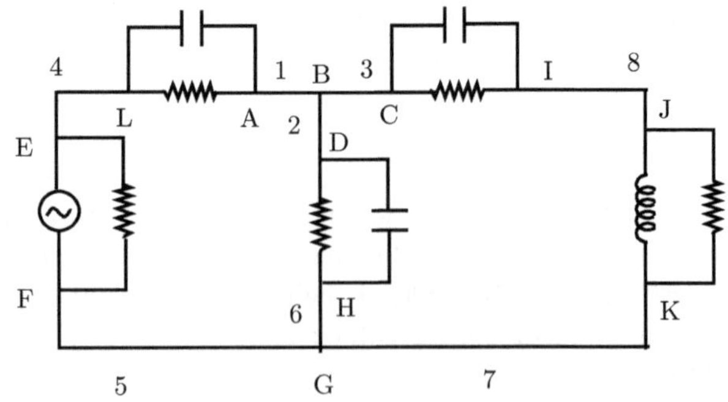

Fig. 6.9 Sample circuit with short nodes and branches

mere inspection, the "short" branches are as a list [1, 2, 3, 4, 5, 6, 7, 8]. A "short" branch can be determined by the fact that it contains only "wire" elements and no components. It should be noted that a "short" branch cannot simply be a branch with zero resistance. For example, if a branch has only an Ammeter which is a zero resistance element, this branch would have a zero resistance but it has a component, i.e., an Ammeter. This is a design error and will be notified to the user who should then correct it as a branch cannot have a zero resistance. If the user wants to have a branch with only an Ammeter, there should be a negligible resistance in the branch. However, if this branch is designated as a "short" branch, it will be deleted from the list of branches and will not appear in the simulator. This is undesirable as the branch contains a component, and since in this case the component is an Ammeter, the user expects to the see the current measured by the Ammeter in the output file. For this reason, the method to check whether a branch is a "short" branch is to check whether it consists of only "wire" elements and not a single component.

The "short" nodes can be clustered together as groups as follows—[A, B, C, D], [I, J], [F, G, H, K], and [E, L]. Since each group of "short" nodes is essentially just one node, one of them becomes the representative node when performing circuit analysis as will be described later. This representative node in a cluster of "short" nodes is usually the first node in the cluster. The determination of "short" nodes may appear trivial but is an iterative process. To begin with, when the "short" branches are determined, the nodes of those branches are collected as node pairs. So, the first iteration will produce these node pairs for the "short" nodes:

```
[A, B], [B, C], [C, D], [I, J], [K, G], [G, H],
        [F, G], [E, L]
```

These pairs of "short" nodes are now iteratively accumulated. Consider any two pairs of "short" nodes above—for example, [A, B] and [B, C] as one pair and [C, D] and [E, L] as another pair. The first pair contains a common node which is node B. This implies that node C is indirectly connected to node A. This concept is used in collecting the nodes together to form "short" node clusters. The list containing the pairs of "short" nodes is repeatedly searched for common nodes. So, in the above example of [A, B] and [B, C], since there is at least one node common between them, these two lists are appended together to form one list [A, B, C] and the second one is deleted. The search algorithm continues to the group [A, B, C] and [C, D], and since again at least one node (node C) is common, the resultant is a single list [A, B, C, D]. In any search iteration through all the node groups, if even a single common node is found between at least two groups of nodes, the search iteration through the node groups is repeated. This is because, this process of appending nodes together may not be completed in a single iteration particularly for more complex circuits as the nodes appear randomly in the list. So even though in the above case, a single iteration might be sufficient to determine all the "short" node clusters, and it would not be advisable to rely on a single iteration to complete the process. The search iteration is terminated only when during an iteration not a single node is found in common between any two node groups. The result of the "short" node search algorithm would be:

```
[A, B, C, D], [I, J], [K, G, H, F], [E, L]
```

With the "short" nodes determined, the "short" branches are deleted from the list of branches. From the above circuit, it is evident that the number of "short" branches is a significant number of all the branches in the circuit. By eliminating these branches from the list of branches, the size of these lists decreases which decreases the burden of the simulator particularly during search algorithms. However, to determine the loops in the circuit, the "short" branches are still essential to determine connectivity. Therefore, the original collection of branches is maintained and branch information to be used for nodal analysis is processed separately as will be described in the next section.

6.5 Connectivity Map for Nodal Analysis

This section will describe how a branch map is created to describe the connectivity of the circuit. This branch map is different from the BranchMap that was generated when the nodes and branches of the circuit were determined. The BranchMap determined in the previous section was a complete map of all branches and nodes in the circuit and is used to determine loops as will be shown in the next section. The branch map discussed in this section decreases the computation time when nodal analysis is performed as the connectivity between nodes can be directly read off from these matrices. Therefore, this branch map is called KCLBranchMap as it is generated principally to ease the computation burden while performing nodal analysis. This section will borrow the concept of "short" nodes described in the previous section and will generate the branch map without "short" branches as these do not play a role in nodal analysis.

To begin with, the list of nodes found previously has to be abridged with respect to the "short" nodes found in the previous section. As stated before, a group of "short" nodes can be considered as a single node and the first node in the group of "short" nodes is chosen as representative of the group. Therefore, while creating this list of nodes called KCLNodes, the check is whether the node exists in any of the "short" node clusters. If it does, it is checked whether the first representative "short" node in the group has been added to KCLNodes. If the first representative "short" node

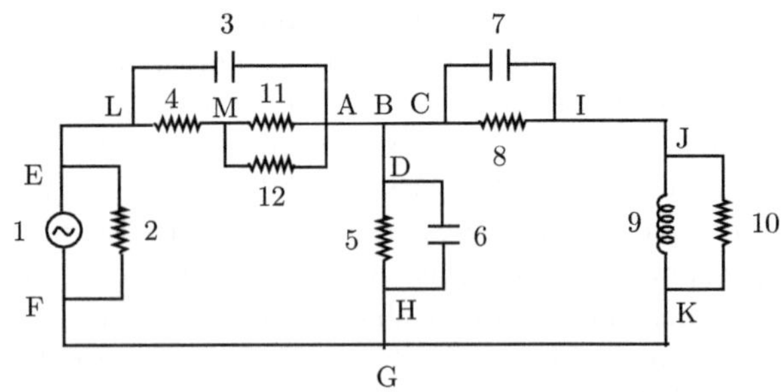

Fig. 6.10 Branch map for nodal analysis

is found in KCLNodes, the "short" node cluster has already been added while if it is not present in KCLNodes, it is added to KCLNodes. If, however, the node being examined is not a "short" node and is in none of the "short" node clusters, the node is added to KCLNodes as it is a regular node. In this manner, the connectivity of the circuit is maintained while removing the "short" branches from the nodal analysis since clusters of "short" nodes are used to check for branch connections. All the nodes in KCLNodes will appear in the nodal analysis to be described in the later chapters, and their voltage will be the result of the analysis. The other "short" nodes will have their voltage equal to the representative first "short" node in the cluster which is a KCLNode and whose voltage has been determined by nodal analysis.

Continuing with building a branch map of the circuit, two such maps are formed. The first map is a detailed map for performing nodal analysis and the second map is a simpler map that just indicates the branches between KCL nodes. Beginning with the simple branch map called BranchesinKCLNodes, this map will be described with the circuit in Fig. 6.10. In the circuit of Fig. 6.10, the branches containing components are numbered 1–12, while the labeling of nodes is with alphabets as in Fig. 6.9. The difference between the circuits is that a new node M is created with branches 11 and 12 between nodes M and A. The purpose of the branch map BranchesinKCLNodes is to indicate which branches are incident at each KCLNode. In the above circuit, the KCLNodes will be:

```
[A, I, K, E, M]
```

Except for node M, all other nodes are present in "short" node clusters and they are the first nodes in their clusters thereby becoming the representative node of that cluster.

For each node in KCLNodes, the simulator searches for all the branches that have this node as one of the terminating nodes or if the terminating node is a "short" node in the same cluster as the KCLNode being considered. For example, the node A will have the following branches incident on it:

```
[3, 5, 6, 7, 8, 11, 12]
```

Only the branches 3, 11, and 12 are directly incident on node A; i.e., one of the terminating nodes of the branches is node A. All other branches are incident at node A through "short" branches. In this case, this can be elaborated as follows. Branches

5 and 6 are incident at node D which is in the "short" node cluster with "short" node A. Branches 7 and 8 are incident at node C which is in the "short" node cluster with "short" node A. Similarly, KCL node I will have branches [7, 8, 9, 10] incident on it, KCL node K will have branches [1, 2, 5, 6, 9, 10] incident on it, KCL node E will have branches [1, 2, 3, 4] incident on it, and KCL node M will have branches [4, 11, 12] incident on it. BranchesinKCLNodes therefore is a list containing lists as follows:

```
[[3, 5, 6, 7, 8, 11, 12], [7, 8, 9, 10],
[1, 2, 5, 6, 9, 10], [1, 2, 3, 4],
[4, 11, 12]]
```

Though BranchesinKCLNodes lets the simulator pick out the branches that connect nodes, it lacks one piece of information that is necessary for nodal analysis—the direction of the branches incident at nodes. Nodal analysis will consider the current entering a node as negative and the current leaving a node as positive. These directions are arbitrary as long as they are consistent; i.e., if the current in branch 3 is assumed to be flowing from L to A, at KCLNode A the current will be negative, while at KCLNode L, the current will be positive. In order to determine the sign of a branch, information about the originating and terminating node as found by the branch search algorithm is used. As stated in the previous section, the branch search algorithm begins at a node and terminates when another node is found. The "direction" of a branch is considered to be from the originating node to the terminating node. This information is combined with BranchesinKCLNodes to form the matrix KCLBranchMap.

KCLBranchMap is a matrix with the row and column dimension being equal to the number of KCLNodes. In the above example, KCLBranchMap will be a 5×5 matrix or in Python terminology will have five lists with each list having five embedded lists. Each element of the matrix will represent a pair of nodes, and for nodes that are not directly connected by a branch, the element will be empty. For example, KCLNode E and I are not directly connected by a branch and therefore [1, 3] and [3, 1] elements of KCLBranchMap will be empty lists. On the other hand, when two KCLNodes are directly connected by one or more branches, the corresponding elements will have a list which in turn will contain the following two lists. The first list will be a list of the branches between the two KCLNodes as it is possible that there will be multiple branches connected between a pair of KCLNodes. The second list will be the directions of each of those branches. For example, in Fig. 6.11, between KCLNode A and KCLNode M, the element [0, 4] will be [[11, 12], [1, −1]]. Branch

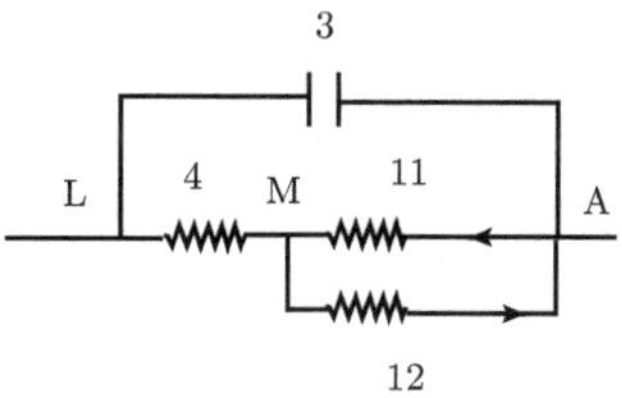

Fig. 6.11 Multiple branches between nodes

11 is originating at KCLNode A which causes the corresponding direction in the second list to be 1; while since branch 12 terminates at KCLNode A, the direction is -1. Conversely, the element [4, 0] of KCLBranchMap between KCLNode M and KCLNode A will be [[11, 12], [-1, 1]] as branch 11 terminates at KCLNode M, while branch 12 originates from it. It is to be noted that these directions of the branches between nodes are arbitrary. Branch 11 could have its direction reversed in which case element [0, 4] will be [[11, 12], [-1, -1]] and element [4, 0] will be [[11, 12], [1, 1]]. The important fact is that the direction lists of the off-diagonal elements [0, 4] and [4, 0] should be opposite in sign— [-1, 1] in the first example and [1, 1] in the second example.

The list BranchesinKCLNodes and KCLBranchMap will be shown to be effective in performing nodal analysis of the circuit and also for simple connectivity checking of the circuit. Another point to be noted is that nodal analysis always has the same structure—the same KCLNodes with the same branches incident on them. However, as will be shown in the next chapter, the loops change with the nonlinear nature of the circuit. This implies that as devices turn on and turn off, the loops that are used to describe the circuit change and therefore, the next chapter will also describe loop manipulation techniques. However, for nodal analysis, only the impedances of the branches will change as the devices change state but the equations for nodal analysis derived from KCLBranchMap remain the same.

6.6 Loops

The simulator uses both loop analysis and nodal analysis to determine the currents in the branches and the voltages at the nodes. The section deals with the determination of loops in a circuit and continues from the list of nodes and list of branches determined in the previous sections. In nodal analysis, there is a current balance (Kirchoff's Current Law) equation for every node in the circuit except for the ground or reference node. Therefore, in nodal analysis, the nodal equations are static and are unique for a given circuit. The only factor that can change is the choice of the reference node if a particular node has not been defined by the user as a reference node. However, in the case of loop analysis, the loops can be defined in numerous ways for a particular circuit.

Considering the example of the circuit in Fig. 6.12, the concept of non-unique loops can be explained further. The three independent loops needed to fully represent the circuit can be drawn in two totally different ways as shown in the two figures. A number of additional permutations are also possible for the user to draw loops. However, the number of independent loop equations will always be $L = B - N + 1$ where B are the number of branches, N are the number of nodes, and L the number of loops (6). In the above circuit, there are six branches (including the one "short" branch) and four nodes (including the two "short" nodes) which result in three independent loops. Loops in addition to these independent loops are linear combinations of the independent loops. For example, in the circuit in Fig. 6.13, the

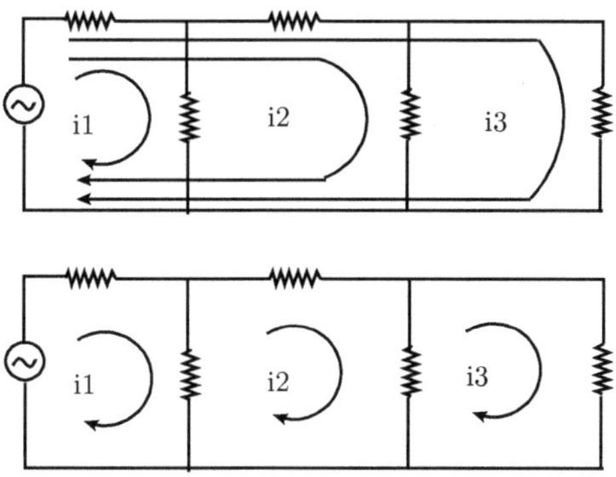

Fig. 6.12 Same circuit different loops

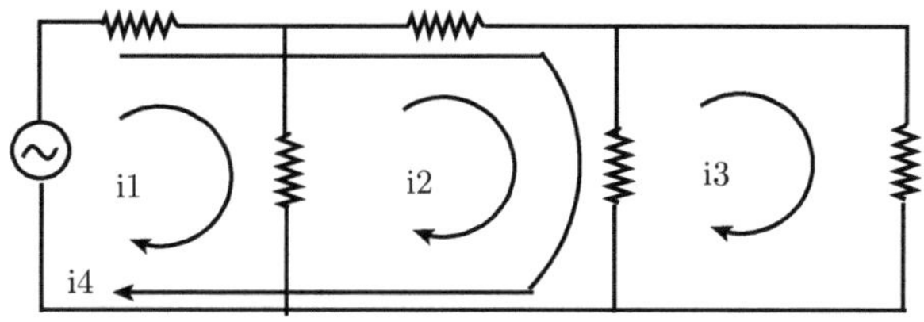

Fig. 6.13 Excessive loops are a linear combination of other loops

loops i1, i2, i3 are independent loops and the additional loop i4 can be seen to be the linear combination i1+i2.

The loop search algorithm begins with the original branch map generated by the branch search algorithm with the "short" nodes and "short" branches being included in this original branch map. This is to reduce the complication in the loop search algorithm as continuity of branches is extremely important while checking for loops. A loop is a closed path in a circuit as can be seen from the above figures. Therefore, the terminating condition is that the loop should end with the originating node. Furthermore, there are several other cases that violate a closed path becoming a loop. For example, a loop cannot encounter a node or a branch more than once. Since a loop is defined as a closed path in a circuit, there is a possibility of a smaller loop embedded inside a larger loop when a node or a branch is encountered twice as in that case a node becomes the originating and terminating node. The process of determining loops is iterative and recursive which is why it was one of the most complicated parts of the circuit simulator.

Consider the above circuit as an example to describe the loop search algorithm. The circuit is repeated in Fig. 6.14 with the nodes marked N1–N4 and the branches marked B1–B6. The result of the previously described algorithms to search for nodes and branches will produce a BranchMap which describes the connectivity between nodes. For the circuit in Fig. 6.14, a sample BranchMap can be described as in Fig. 6.15. This BranchMap of Fig. 6.15 appears as lists embedded within a list. The loop finder algorithm uses the above BranchMap to determine loops. Before describing the algorithm, a few loops will be "drawn" in this BranchMap to illustrate the concept.

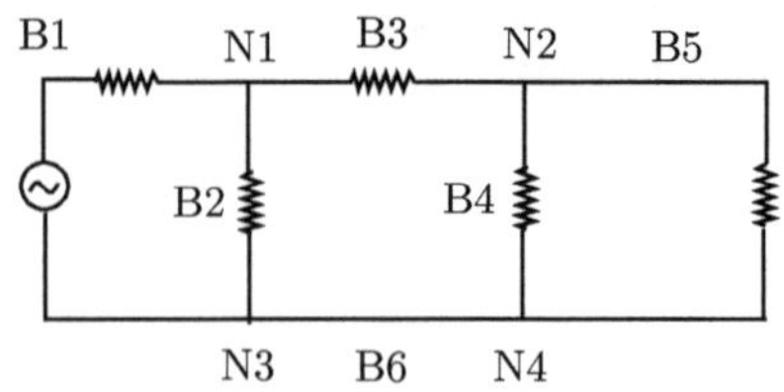

Fig. 6.14 Formulation of the loop search algorithm

Fig. 6.15 Branch map used in loop search

	N1	N2	N3	N4
N1	X	B3	B1,B2	X
N2	B3	X	X	B4,B5
N3	B1,B2	X	X	B6
N4	X	B4,B5	B6	X

Fig. 6.16 Loop found in the branch map

	N1	N2	N3	N4
N1	X	B3 →	B1,B2	X
N2	B3	X	X	B4,B5
N3	B1,B2	X	X	B6
N4	X	B4,B5 ←	B6	X

Consider the loop described in Fig. 6.16 with the following branches—B3-B2-B6-B4-B3. The actual loop is shown in the circuit diagram in Fig. 6.17 and can be seen to be a genuine loop. However, the arrow between the branches in BranchMap cannot be random. For example, an arrow has been drawn from B3 to B2. However, this is possible because B3 and B2 share a common node which in this case is N1. Therefore, while extending the loop by a branch, it is checked whether the new branch has a common node with the latest branch in the developing loop. This is how the loop search can be described.

1. The loop starts with the first branch B3.
2. The algorithm checks for the next branch and finds branch B2 with node N1 to be common and adds it to the loop which becomes B3-B2.
3. The next branch is branch B6 with node N3 in common with the last branch B2 in the developing loop. So, the loop is now B3-B2-B6.

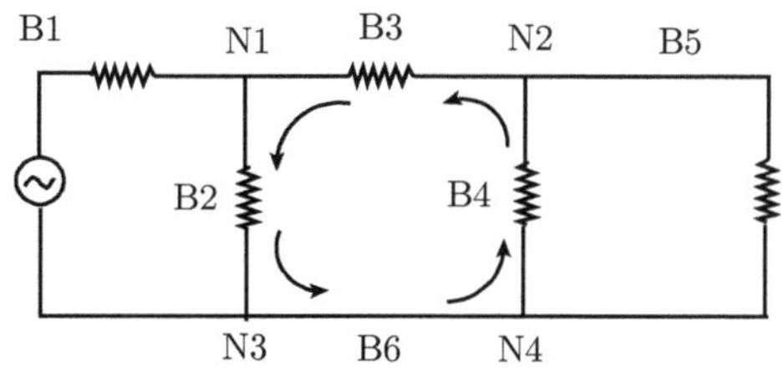

Fig. 6.17 Actual loop in the circuit

4. The next branch B4 has node N4 in common with the last branch B6 in the developing loop. So, the loop is now B3-B2-B6-B4.
5. The next branch B3 is the originating branch, and this is where the loop closes to be B3-B2-B6-B4-B3.

The above process will be generalized to be able to search for any loop as described below.

To begin with, there are two types of loops that the loop algorithm must search for. The first is the parallel loops between a pair of nodes which in the above example would be B1-B2 and B4-B5. The second is all other types of loops that have more than two nodes. There is no limit to the number of branches that can be connected between two nodes, and therefore, for B branches connected between two nodes, the number of loops would be $B-N+1=B-1$. By forming loops between parallel branches connected between nodes, the burden of the loop algorithm decreases as the number of loops that it needs to search for is only those with more than two nodes. In most power electronics circuits, it is quite normal to have a large number of parallel loops between nodes. For example, voltage measurement would imply the connection of a Voltmeter across a pair of nodes. In many cases, the output of a power converter would be available at a filter capacitor with a resistor and a Voltmeter connected across it. In the simple example above, the circuit has two loops between parallel nodes B1-B2 and B4-B5 but has only one loop with more than two nodes.

To generalize the algorithm for the loop with more than two nodes, we examine every step 1–5 of the loop algorithm listed for the sample circuit of Fig. 6.17. The first element can be any random branch between any two nodes. To ensure that no loops are missed, the search algorithm will begin a new search with a branch between every pair of nodes. Such a search technique may be a burden as many of the loops will be duplicates of loops already found but as will be shown later, it will ensure that the collection of loops will always be linearly independent. The BranchMap for a circuit with Ni nodes and the algorithm for creating new searches with the loop algorithm is shown in Fig. 6.18.

In BranchMap, after choosing a starting branch, the search algorithm has to proceed to the next branch. To convert this concept that is visually obvious into an algorithm, the structure of a loop needs to be understood. A loop is a collection of branches such that they form a close path while ensuring that a branch is not repeated or a node is not repeated. These two checks are repeatedly performed as the algorithm executes. However, to reduce the computation time of the algorithm,

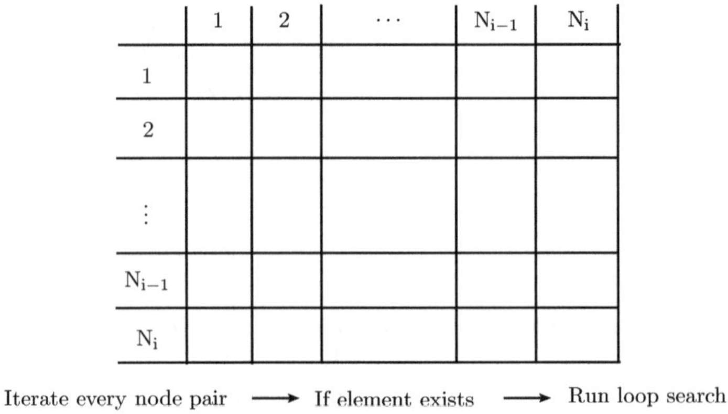

Fig. 6.18 The scope of the loop search algorithm

Fig. 6.19 Beginning of a loop search iteration

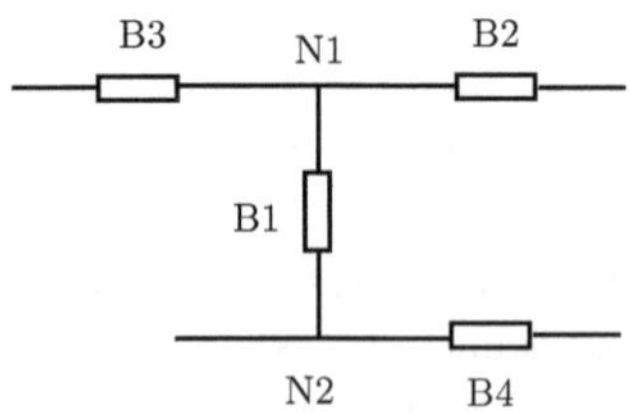

branches that will outrightly violate the repetition of a branch or a node will not be added. Let us consider the following circuit segment in Fig. 6.19 where branch B1 is the first branch that the loop algorithm starts with. Since B1 is the first branch, the loop algorithm can add branches B2, B3, or B4 as the next branch. With respect to BranchMap of Fig. 6.20, the choice can be described with respect to the nodes of the branch B1. The branch B1 is between nodes N1 and N2 as can be seen from the BranchMap. Due to BranchMap being a symmetrical matrix, every branch appears twice, for example branch B1 as the node pair N1, N2 occurs twice—as (N1, N2) and (N2, N1). From the circuit diagram, branches B2 and B3 are also incident at node N1. The circuit diagram does not specify the other terminal nodes of B2 and B3. In BranchMap, these two branches appear in the row N1 and the column N1. Branch B4 is incident at node N2 and therefore appears in row N2 and column N2.

When starting a loop with branch B1, there are three possible branches that can be added—branch B2, branch B3, or branch B4. By moving along the row from branch B1, the branches that can be added are branch B2 or branch B3. Moving along the row implies looking for branches that are incident at the node corresponding to the row which in this case is N1. Moving along the column implies looking for branches that are incident at the node corresponding to the column which in this case is N2. The loop search algorithm can proceed in only one direction and with only one branch to ensure that the violation of adding a node more than once does not encounter. As a default, the starting direction from the originating branch is taken to be horizontal or

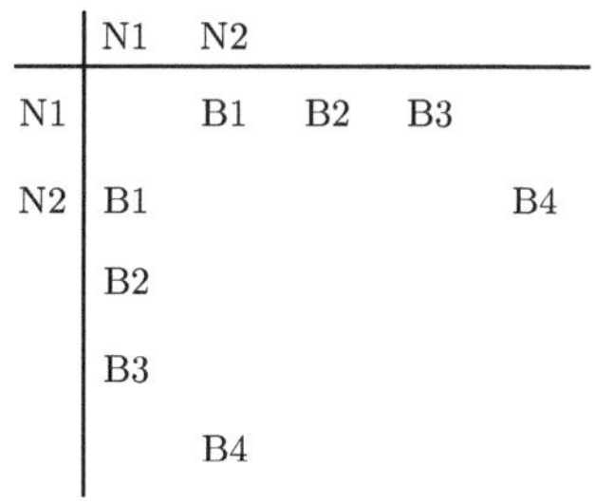

Fig. 6.20 Addition of a branch in BranchMap

Fig. 6.21 Search direction from a branch

	N1	N2	N3	N4	N5	N6	N7
N1		B1	B2	B3			
N2	B1						
N3	B2				B5	B7	
N4	B3						B6
N5			B5				
N6			B7				
N7				B6			

along the row and the branches to be added can be branch B2 or branch B3. As stated before, the loop algorithm attempts to close the loop with any number of branch combinations as it is impossible to know until a terminating condition has reached whether a particular search iteration would result in a valid loop. Therefore, both branch B2 and branch B3 are added separately to create two temporary loop lists [B1, B2] and [B1, B3].

The next step in the loop search algorithm is the addition of a random branch. Let us consider the continuation of the previous example where now B5, B6, and B7 are available to be added as can be seen from the following BranchMap in Fig. 6.21. It should be noted that the loop search has already bifurcated the probable loops starting with branch B1 into [B1, B2] and [B1, B3]. Therefore, the continuation of the loop search is from branches B2 and B3 for their respective incomplete loops. Figure 6.21 below shows the branch map with these branches B5, B6, and B7. Branch B5 is between nodes N3 and N5, branch B6 is between nodes N4 and N7, and branch B7 is between nodes N3 and N6. At this point, the circuit diagram has been deliberately deferred to show the position of the branches on the BranchMap and how the algorithm adds branches to the loop list.

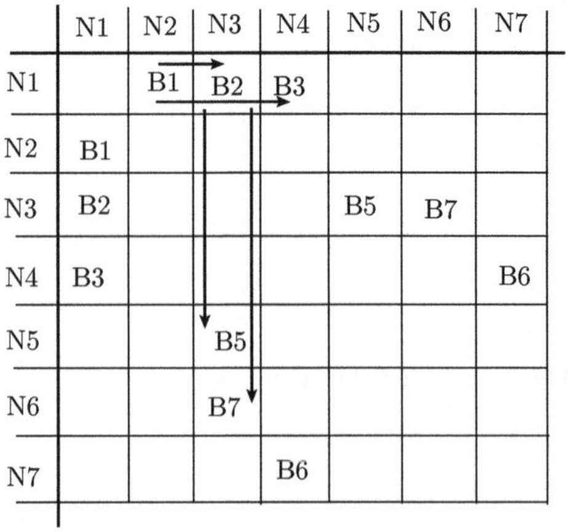

Fig. 6.22 Change of search direction from a branch

To examine how the loop algorithm will proceed, we will revert back to the previous step when branches B2 and B3 were added after the starting branch B1. After choosing the starting branch B1, the algorithm moved along the row corresponding to node N1 and found branches B2 and B3 that were also incident on node N1. This is shown by the two arrows starting from branch B1 in Fig. 6.21. The loop algorithm is now looking for branches to add after branches B2 and B3. Let us consider the temporary loop list [B1, B2]. Since B1 and B2 have node N1 in common, the next branch will have to be incident on the other node of branch B2 which in this case is node N3. At node N3, besides branch B2, there are incident on it branches B5 and B7. From Fig. 6.22, the search for branches incident on node N3 can be seen as moving along the column corresponding to node N3. As the direction of arriving at branch B2 from the previous branch B1 was along the row, the direction of search from B2 will be along the column. This is shown by arrows in Fig. 6.22.

In a practical electrical circuit, a self-loop is not possible. A self-loop can exist in a purely mathematical construct but cannot be visualized in an implementable circuit. Therefore, in BranchMap, every branch exists in an element whose row and column are not the same. The diagonal elements of BranchMap are always empty and contain no branches. The loop algorithm always changes search directions after every iteration of adding branches. From Fig. 6.22 of the BranchMap, when a branch is arrived at from a direction say along a row, it has been added when "looking" for branches from the node corresponding to the row. When the loop algorithm needs to continue the search from this branch, it must look from the other node of the branch which in this case is the node corresponding to the column. Therefore, the loop search algorithm now proceeds along the column. Exactly the opposite would have occurred if the loop algorithm arrived at the branch while moving along a column. The loop algorithm would then have proceeded from the branch along the row. In this manner, the loop algorithm always changes its direction after adding branches at every iteration.

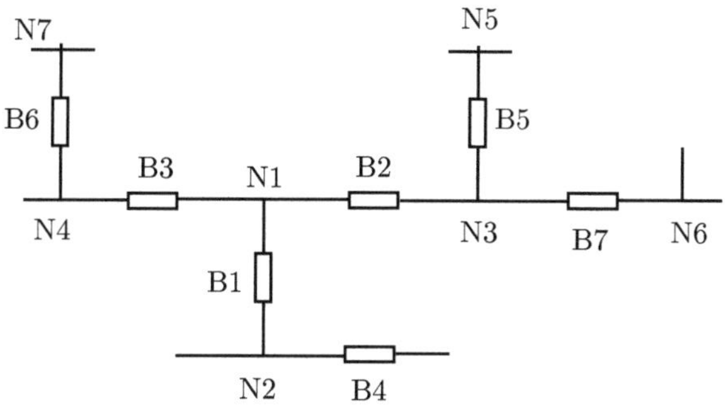

Fig. 6.23 Actual circuit with respect to the BranchMap

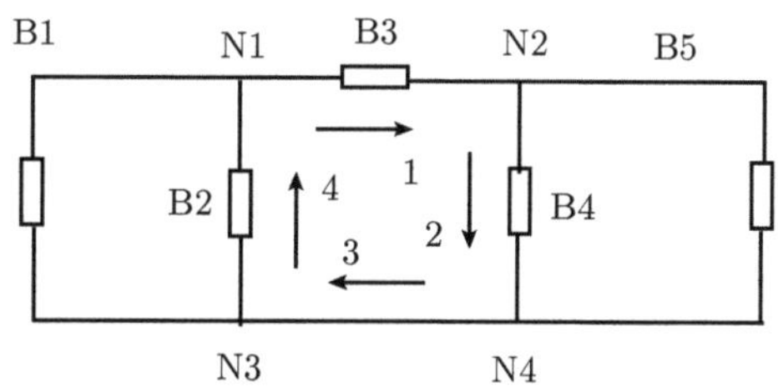

Fig. 6.24 A successful loop in a circuit

With the above explanation, Fig. 6.23 will show how the above BranchMap translates into the following actual circuit. Any circuit can be represented with such a BranchMap matrix, and therefore, the above loop algorithm can be used for any circuit. As stated before, the loop algorithm will examine every possible method of finding closed loops which means the temporary loop list [B1, B2] will be bifurcated to add both branches B5 and B7 as [B1, B2, B5] and [B1, B2, B7]. Furthermore, branch B6 will be added to the list [B1, B3] to become [B1, B3, B6]. As a result, the total number of loop lists starting with branch B1 that have been generated by the loop search algorithm so far is [B1, B2, B5], [B1, B2, B7], [B1, B3, B6]. The number of possible loops can therefore increase with every branch added, and for complex circuits, these temporary incomplete loops can grow to a fairly large number. However, very few of these lists will result in successful loops and a large number of them will be deleted when the loop algorithm aborts. This will be described below.

The successful closing of a loop can be described with the circuit in Fig. 6.24. The loop consists of branches B3-B4-B2 including a short branch encountered in the sequence 1-2-3-4 as shown above. The first branch is B3, and the exit node from which the branches are added by the loop algorithm is N2. When the final branch B2 is added, one of its nodes N1 is also a node of the starting branch B3 and this implies successful closure of the loop. Therefore, in general, when a branch is added by the loop algorithm which has one of its nodes to be the same as the originating node of the starting branch, the loop is said to have successfully terminated. The concept of the originating node of the starting branch needs more description. When the loop search algorithm begins with the starting branch, which in this case is branch B3, the position of branch B3 in BranchMap is (N2, N1). This is due to the fact that the search direction from the starting branch is along the row and therefore this row will be the one corresponding to node N2. Node N2 is the exiting node of the starting branch, while node N1 is the originating node. Therefore, the loop will be successfully closed only when the terminating branch has one of its nodes to be same as the originating node N1.

The loop search algorithm attempts to close a loop by adding every possible branch. This results in repeated bifurcations that cause the collection of temporary possible loops that start with a branch to increase. However, only a few of these loops will be closed, and many of them will be deleted. This section describes the cases when loops are deleted as being ineligible loops. The simplest case of a loop becoming ineligible is when it attempts to add a branch which has already been added before. When this simple check fails, it means the loop is traversing the same branch and therefore is no longer a single closed path. An extension of this case is when the loop search algorithm arrives at a branch and finds that all other branches incident at the exit node of this branch have already been added in the loop. Therefore, there is no possible continuation of the loop and the loop is discarded. Another case of failure is when a node is repeated but this is in addition to the terminating condition. An example will be the circuit and the loop shown in Fig. 6.25. The circuit shows the loop 1-2-3-4-5-6-7. It meets the terminating condition of a loop and also follows the logical sequence of branches been added to the loop until it successfully closes. However, this is not a valid loop as it contains a smaller loop embedded within it which is 2-3-4. The reason is that the node N2 is encountered twice. When a node is encountered twice, it means a loop has been formed. However, a single closed path is formed only when the node is encountered twice in the terminating condition. When node N2 is encountered for the second time, the terminating condition is not met as branch 4 does not have one of its nodes to be the same as the originating node N1 of the starting branch 1. Therefore, the loop does not end here but continues on to add branches 6 and 7 before closing successfully. When a node is encountered twice but not in the terminating condition, it almost always means an embedded loop will be formed. The check for this condition is to exclude the originating node of the starting branch and check if all other nodes of the other branches in the loop have been encountered only once. If any of these nodes are encountered more than once, the loop is discarded.

The loop search algorithm will at the end check if any of the loops are repeats of each other. This is always possible as the search algorithm starts with every branch in BranchMap and attempts to find closed loops. A loop could be repeated but in a different sequence. When loops are repeated, the first loop is retained while all subsequent repetitions are discarded. There is a more complicated case of redundancy which cannot be identified at this stage and will be described in the next chapter. As described in the beginning of the section on loops, it is possible to write loops in a number of different ways. Moreover, it is also possible to write additional loops to describe a circuit. A circuit will have B-N+1 number of linearly independent loops that are needed to accurately describe it. Additional loops will be a linear combination of these linearly independent loops. However, these additional loops may not be repetitions of the linearly independent loops and therefore will not be

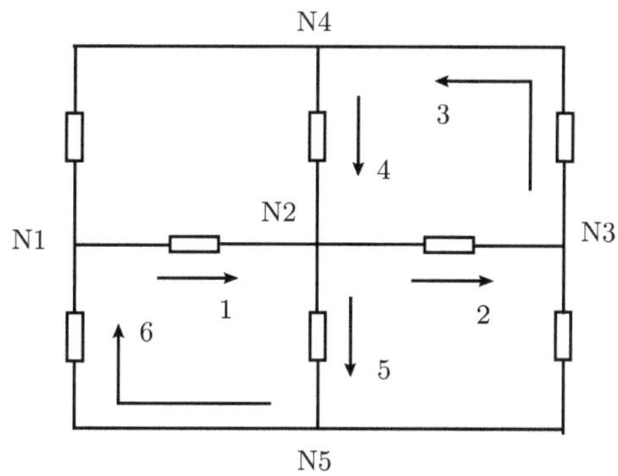

Fig. 6.25 A failed loop search

identified by the repetition check. They can only be identified when elementary operations are performed on the sets of loop equations. The elementary operations are in the form of row operations and result in the additional equations becoming null equations which can then be discarded.

6.7 Loop Map

The loop search algorithm described above will return a list of completed loops. Each completed loop as described above is a list of branches with each branch represented by a pair of nodes that are the terminal nodes of the branch. However, this list omits an important information with respect to the branches. This is the direction in which the branches appear in the loops. The branches are assigned arbitrary directions depending on how the search algorithm determines them, i.e., in terms of starting node and ending node. The current in a branch is considered positive when it flows from the starting node to the ending node. Any voltage source in a branch is considered positive when the positive polarity of the source is closer to the starting node of the branch. As a result, when performing loop analysis which will be described in the next chapter, the direction in which a branch appears is important and must be captured.

Let us reconsider the circuit of Fig. 6.26 presented before in this chapter but this time with directions on the branches. The loop consists of the branches B3-B4-B2 with one short branch. The direction of the loop is marked by the inner arrows 1-2-3-4 which is how the loop search algorithm adds the branches. The directions of branches are indicated next to the branch labels. Branch B3 is from node N2 to N1, branch B4 is from N4 to N2, and branch B2 is from N3 to N1. Therefore, when the loop search algorithm adds the branches, branch B2 is along the direction of the loop while branches B3 and B4 are against the direction of the loop. To be able to add the information about the direction of a branch with the loop, the exit node from a branch is of importance. For example, when the loop starts with branch B3, the loop

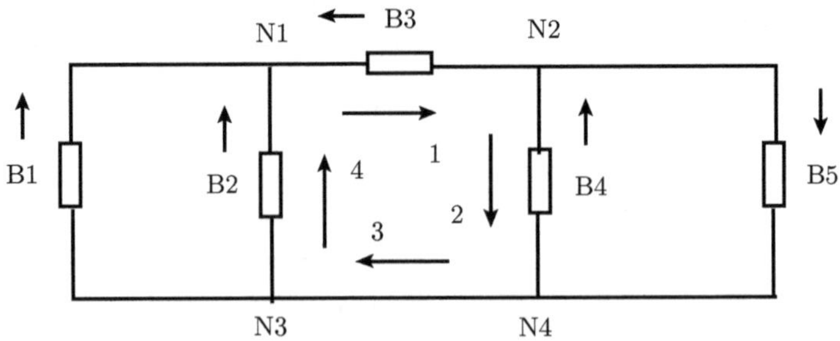

Fig. 6.26 A loop with respect to branches with their directions

search algorithm has added the element (N1, N2) from BranchMap. After this, the search algorithm looks for branches incident at node N2. Therefore, node N2 is the exit node with respect to branch B3. However, node N2 is the originating node of branch B3 which is from node N2 to node N1. This is an indication of a branch being in the direction opposite to that of the loop. Similarly, at exit node N2, the search algorithm finds branch B4 incident on it. However, node N2 is the terminating node of branch B4 which makes the direction of branch B4 to be opposite to the direction of the loop. On the other hand, let us consider the case when the loop algorithm adds branch B2. After adding the short branch between node N4 and node N3, the search algorithm looks at branches incident at node N3. Node N3 being the exit node at this point, branch B2 is found incident at node N3 and this node N3 is the originating node of branch B2. Therefore, branch B2 is in the same direction as the direction of the loop.

The loop search algorithm adds the loop described above as a set of node pairs. The loop indicated in Fig. 6.27 will be:

```
[N1, N2], [N2, N4], [N4, N3], [N3, N1] ->
                 B3-reverse, B4-reverse, B2-forward
```

For the circuit shown, this is the only loop that has more than two nodes. The other loops are merely loops comprising parallel branches between nodes. The other two branches will be:

```
[N1, N3], [N3, N1] -> B2-reverse, B1-forward
[N2, N4], [N4, N2] -> B4-reverse, B5-reverse
```

As stated previously, the loops comprising parallel branches between pairs of nodes will be treated separately to reduce the burden of the loop search algorithm. If any element in BranchMap contains more than one branch, loops can be written between the branches in that element. The method of determining the direction of the branch within such a loop remains the same.

In order to combine the information of branches and their directions above into a single presentable format, we introduce the concept of the LoopMap. Similar to the BranchMap, the LoopMap allows quick visualization of the branches and their directions in each loop of the circuit as shown in Fig. 6.27 for the same circuit in

Fig. 6.27 LoopMap representing the loops in the circuit

	B1	B2	B3	B4	B5
L1		FW	RV	RV	
L2	FW	RV			
L3				RV	RV

Fig. 6.26. This matrix provides both pieces of information for all the loops in a circuit—the branches present in the loop and the direction of the branch with respect to the loop. The “FW” and “RV” in the map mean the branches are in the direction (forward) and against the direction (reverse) of the loop. This LoopMap will form the basis for the loop analysis to be described in the next chapter. LoopMap can be used for generating the matrices used for solving differential equations iteratively to update the loop and branch currents. Moreover, LoopMap also provides the facility to perform loop manipulations as will be described in detail in the next chapter.

6.8 Conclusions

This chapter presents circuit data in a format that can be used by the circuit solvers that will be described in Chaps. 7 and 8. The circuit solver consists of solving linear equations that may be either purely algebraic or differential equations. Loop analysis is a combination of solving ordinary differential equations and purely algebraic equations. Nodal analysis involves solving purely algebraic equations. As would be evident, loop analysis performed at a particular instant of time requires information of all the loops in the circuit at that instant. This information is made available through the LoopMap described in the last section. Nodal analysis requires information of nodes and branches which is made available in the form of the KCLBranchMap.

During the simulation process, information flows both ways between the circuit data and the circuit solver. The circuit solver generates the matrices for solving equations based on the latest state of the circuit as captured in the branches and the loops. For example, the branches of the circuit will contain information of the total resistance, total inductance, current, and the total voltage in the branch. As time progresses, if a branch has an ac voltage source, the total voltage in the branch will change. If the branch has a variable resistance, the value of the total resistance of the branch will change if the variable resistor is altered by control code. The value of the current in a branch could change due to a change in the branch or due to changes in other branches of the circuit. In this manner, at the end of every simulation iteration, all branch data will be updated to capture the latest state of the circuit.

When the circuit solver is launched, the data in the branches are the foundation for generating the equations for both loop and nodal analyses. For loop analysis, the LoopMap is updated based on the state of the circuit. As shown, LoopMap describes loops in terms of the branches present in a loop and the direction of the branch with respect to the loop. Therefore, branch data can be used to generate equations for loop analysis from LoopMap. In a similar manner, nodal analysis requires information of the nodes in the circuit and the branches that connect these nodes. The difference between loop analysis and nodal analysis lies in the fact that loops are not unique and can be written in a number of different ways for a given circuit, but the connectivity map for nodal analysis is always unique. This connectivity map for nodal analysis is captured in KCLBranchMap which contains information about the branches between the nodes and their direction. The equations for nodal analysis are generated by using branch data with KCLBranchMap.

This chapter has described how nodes, branches, and loops are determined from the circuit schematic that was described in Chap. 3. The concept of branches, nodes, and loops has been described without reference to the circuit components but merely with the presence of an element— a "wire", a component, or a jump label. Therefore, a node can only be a "wire", a branch is a collection of the above elements, and a loop is a collection of branches. Since a branch is a collection of elements, it becomes a collection of all the objects associated with circuit components as described in Chap. 3. Therefore, branch data are in turn generated by extracting data from the objects corresponding to the circuit components. And conversely, after each simulation iteration the objects corresponding to the circuit components are updated based on branch data which in turn is updated from the results from loop and nodal analyses.

Besides being the foundation for performing circuit analysis, the structures LoopMap and KCLBranchMap provide a wealth of information about the circuit. As will be shown in the next chapter, LoopMap can be altered by performing loop manipulations and this is a technique which has to be used when the circuit is nonlinear. The loops in such a case of a nonlinear circuit provide information about the paths for the currents to flow, and a loop is by definition a closed path in a circuit. Therefore, going through LoopMap or may be just segments of the LoopMap are an effective way to debug a circuit by determining the paths through which the current flows in the circuit. The source code contains some functions that allow a user to display either all or some of the loops in LoopMap. These functions were originally designed to debug the simulator during the development process but now serve as a learning tool while simulating large and interconnected circuits.

Chapter 7
Circuit Analysis—Loop Analysis

Abstract This chapter describes how loop analysis is performed in the simulator. The chapter describes how the matrix equation for performing loop analysis is generated from the LoopMap described in Chap. 6. A brief description is provided about how the matrices in this equation are transformed using row operations such that they can be solved by using numerical integration techniques. The chapter describes how loop currents and branch currents in the circuit can be mapped which allows for calculation of branch currents from loop currents and vice versa. The chapter describes with an example how time constants of branches of the circuit can make the simulation unstable and introduces the concept of a stiff loop. By providing a sample circuit and its corresponding LoopMap, the chapter describes the need to isolate stiff loops so as to be able to simulate a circuit. With this example, the concept of loop manipulations is described and with advanced examples, the effectiveness of the procedure is described. The chapter describes the limitation of loop analysis with another set of examples and therefore the need for nodal analysis.

Keywords Loop analysis · Ordinary differential equations · Elementary operations · Numerical methods · Solution stability · Stiff systems · Time constants · Loop manipulations

7.1 Introduction

Chapter 6 described the process of determining nodes, branches, and loops in a circuit from the circuit schematic. As stated in the previous chapter, the LoopMap and the KCLBranchMap are the foundations for performing loop analysis and nodal analysis. This chapter will describe loop analysis leaving the next chapter to describe nodal analysis. This chapter will also describe the technique of loop manipulations which alter the loops in LoopMap. Loop manipulations become extremely important in simulating nonlinear circuits which are mainly described in the next chapter, but for which the foundations are laid toward the end of this chapter.

Circuit analysis can be performed using loop analysis and/or nodal analysis [6]. As stated in the previous chapter, the loops that describe a circuit can be written

S. V. Iyer, *Simulating Nonlinear Circuits with Python Power Electronics*,
https://doi.org/10.1007/978-3-319-73984-7_7

in a number of ways. This is a disadvantage of loop analysis, and as for nonlinear circuits, the loops need to be rewritten every time there is a change in the circuit as the closed paths followed by currents change. In contrast, nodal analysis uses the connectivity information between nodes in the circuit which for a given circuit is always the same. It is only the parameters of the branches that change, resulting in different nodal equations. Therefore, in most cases, performing nodal analysis reduces the computational burden as there is no need to recompute KCLBranchMap and only the matrices for the equations have to be generated.

However, power electronic circuits differ from most analog circuits for which most circuit simulators are targeted. Power electronic circuits very often have several inductors, as the only way to control a nonlinear circuit is by using energy storage devices which are inductors and capacitors. In contrast, most other analog circuits are comprised of resistor and capacitors and rarely have inductors and if they do, quite rarely is the objective being the need to control a circuit with the inductor as an energy storage element. When a circuit contains a large number of inductors, solving the circuit using nodal analysis is tricky as these inductors have the property of not letting the current through them change instantaneously. In other words, to be able to solve a circuit with inductors, the $\frac{di}{dt}$ through the inductor needs to be determined. Performing nodal analysis typically results in the nodal voltages being determined with respect to the resistances, voltages sources, and current sources in the circuit. The question is how is an inductor now to be modeled. As a voltage source $L\frac{di}{dt}$ or as a current source?

It is always possible to approximate the inductor in nodal analysis with respect to the above question and be able to solve it. However, given the nature of the inductor, loop analysis is a much more convenient manner of solving a circuit with inductors. To elaborate, loop analysis sums up the voltages in a closed path—voltage drops across resistors and inductors and the voltage sources. The voltage drop across the inductor is $L\frac{di}{dt}$ and since the variable to be calculated in loop analysis is the loop current, this readily allows us to solve a differential equation based on the $\frac{di}{dt}$ of the inductors. The output of loop analysis is the loop currents from which not only inductor currents but all the currents in the circuit can be determined. In contrast, nodal analysis will produce the node voltages in the circuit, from which the currents in branches having inductors will then need to be calculated.

The method of performing loop analysis will be described in detail in this chapter. The method of generating equations from LoopMap will be described with sample circuits. Moreover, the concept of stiff systems will be introduced with a sample circuit and how simulating one requires amendment of loop equations. The concept of stiff systems is then the basis for introducing the concept of loop manipulations which is again described with sample circuits that have stiff branches. Finally, the limitation of loop analysis is described which is then overcome with nodal analysis in the next chapter.

7.2 Matrices for Loop Analysis

The loop analysis on a circuit is performed by writing the Kirchoff Voltage Law (KVL) equations as differential equations. In order to solve the equations, they are represented as matrices and solved using matrix manipulation techniques. This section will describe how the matrix equations are written for any circuit once a LoopMatrix has been derived as described in the previous section. Consider the circuit in Fig. 7.1 which is a more detailed version of the circuit presented before. The loops L1, L2, and L3 and the branches are indicated with their directions. For the circuit in Fig. 7.1, the LoopMap will be as shown in Fig. 7.2. In this example, the components are used in the branch columns instead of only the branch symbols as the values of the components will be used in the loop analysis to follow. For the loops drawn for the above circuit, the loops can be written in the form of the equations below. Let us consider the voltage source V1 to be a sinusoidal voltage waveform with an instantaneous value of V1 and the loops 1, 2, and 3 to have the loop currents I1, I2, and I3.

$$V_1 - (R_1 + R_2)I_1 + R_2 I_2 = 0 \tag{7.1}$$

$$R_2 I_1 - (R_2 + R_3 + R_4)I_2 - L_1\frac{dI_2}{dt} + R_4 I_3 = 0 \tag{7.2}$$

$$R_4 I_2 - (R_4 + R_5)I_3 - L_2\frac{dI_3}{dt} = 0 \tag{7.3}$$

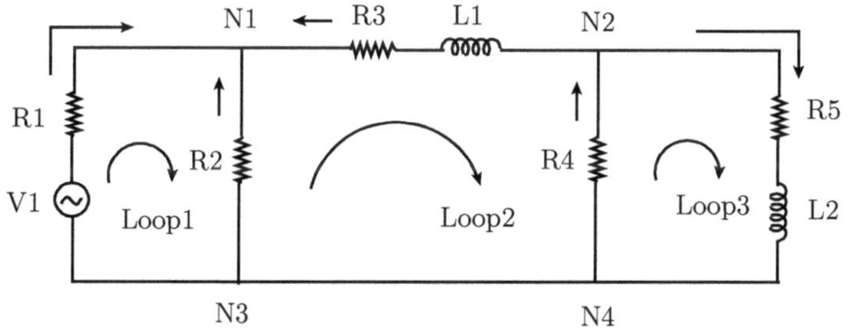

Fig. 7.1 Loops in a circuit

Fig. 7.2 LoopMap for the circuit

	{V1,R1}	R2	{R3,L1}	R4	{R5,L2}
Loop1	FW	RV			
Loop2		FW	RV	RV	
Loop3				FW	FW

The above equations are fairly easy to verify. For a circuit as simple as the one above, the equations can be written individually and solved simultaneously. However, as the circuit increases in size and the number of loops increases, the only convenient way to solve such a system of equations is in the form of matrices. The above set of equations can be written in the following matrix form:

$$\mathbf{E}\frac{d\mathbf{x}}{dt} - \mathbf{A}\mathbf{x} = \mathbf{B}\mathbf{u} \tag{7.4}$$

In the above matrix equation, the vector **x** is the vector of variables which in this case are the loop currents I1, I2, and I3. $d\mathbf{x}/dt$ denotes the differential or derivative of these variables with respect to time. The vector **u** is the collection of all the sources which in this case is V1. **E**, **A**, and **B** are matrices that can be written as follows:

$$\mathbf{A} = \begin{bmatrix} R_1 + R_2 & -R_2 & 0 \\ -R_2 & R_2 + R_3 + R_4 & -R_4 \\ 0 & -R_4 & R_4 + R_5 \end{bmatrix} \tag{7.5}$$

$$\mathbf{E} = \begin{bmatrix} 0 & 0 & 0 \\ 0 & L_1 & 0 \\ 0 & 0 & L_2 \end{bmatrix} \tag{7.6}$$

$$\mathbf{B} = \begin{bmatrix} 1 \\ 0 \\ 0 \end{bmatrix} \quad \mathbf{U} = \begin{bmatrix} V_1 \end{bmatrix} \quad \mathbf{x} = \begin{bmatrix} I_1 \\ I_2 \\ I_3 \end{bmatrix} \tag{7.7}$$

A quick inspection of the matrices will verify that the matrices will form the equations. The next step is to develop an algorithm that can generate the above matrices for any LoopMap representing any circuit.

An examination of matrices **E** and **A** will show two characteristics. First, these matrices are symmetrical. Second, the diagonal elements of these matrices will never be negative. A detailed explanation is as follows. The off-diagonal elements of the matrices are an indication of how the loops interact with each other. Examine the first KVL equation. Loop 1 and Loop 2 have resistor R2 in common. Additionally, these loops oppose each other when they pass through R2. Therefore, the interaction between Loop 1 and Loop 2 is -R2. Therefore, the element (1, 2) and (2, 1) of the matrix **A** is -R2. In the circuit, the branches between the loops do not have any inductance and therefore the matrix **E** has all off-diagonal elements to be zero. However, the loops for the above circuit could be drawn in a different manner and the matrix E would also have off-diagonal elements in that case. The reader is encouraged to examine this. In any case, since the off-diagonal elements of matrices **A** and **E** are the interactions between the loops, these matrices will be symmetrical. However, the signs of the off-diagonal elements can be positive or negative. In the above circuit, Loop 1 and Loop 2 oppose each other over branch R2 while Loop 2 and Loop 3 oppose each other over branch R4. For this reason, all off-diagonal elements are negative. However, if the direction of Loop 3 was reversed and Loop 2 and Loop 3 were in the same direction over branch R4, the elements (2, 3) and (3, 2) of matrix **A** would be +R4. Additionally, two loops could interact over a number of branches. In that case, the off-diagonal elements of matrices **A** and **E** will contain the sum total of the resistances and inductances of the branches. To summarize, if two loops m and n oppose each other over branches with total resistance R and total inductance L, the element (m, n) and (n, m) of matrix **A** will be -R and that of matrix **E** will be -L. Conversely, if two loops m and n assist each other over branches with total resistance R and total inductance L, the element (m, n) and (n, m) of matrix **A** will be R and that of matrix **E** will be L.

Now the second characteristic—the diagonal element will never be negative. The diagonal elements in matrices **A** and **E** are the sum total of the resistances and inductances in the loops. When applying KVL, a voltage source in a loop is positive when the loop progresses from the negative terminal to the positive terminal. Similarly, the voltage drop across a resistance or an inductance is negative when progressing against the direction of the loop. Therefore, even for diagonal elements, one would expect a possibility of both positive and negative terms. However, let us assume that a loop is always traversed in the direction of the current indicated. Therefore, voltage drops across resistances and inductances will always be negative as per KVL. When expressed as matrix equations, the KVL equations will always result in these diagonal terms of the matrices A and E to be positive. The reader is invited to write KVL equations for the above circuit or for that matter any other circuit and examine this fact. The sign appears only in the voltage sources as these can be encountered in any direction with respect to the loop current. For example, in the circuit above, Loop 1 has the voltage source V1 which is taken as positive since the loop progresses from the negative terminal to the positive terminal and element 1 of matrix B is +1. The diagonal elements (1, 1) of matrix **A** and **E** are positive. However, if the direction of Loop 1 were to be reversed, the loop would progress from positive terminal to negative terminal of V1 and therefore this source would appear as -V1 with element 1 of matrix B becoming -1. However, even in this case, element (1,1) of matrices **A** and **E** will continue to remain positive.

Now to describe how the matrix **B** is generated for a circuit. The matrix **B** relates the voltage sources in the circuit to the loop equations. Therefore, to begin with, a list of all the voltage sources in the circuit is prepared. Let us consider a more complex circuit with more than one voltage source to describe this process. The circuit in Fig. 7.3 contains 8 branches and 5 nodes and therefore will have 8-5+1=4 independent loops. These loops are drawn above by mere visual inspection, but however, they could very well be conceived by the loop search algorithm described in the previous chapter. The above circuit has three voltage sources V1, V2, and V3. The input vector **u** in the matrix equation will be a collection of these three voltage sources and will therefore be

$$\mathbf{u} = \begin{bmatrix} V_1 \\ V_2 \\ V_3 \end{bmatrix} \tag{7.8}$$

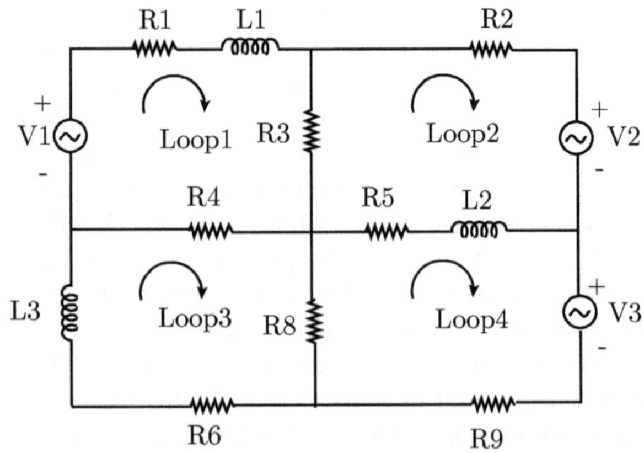

Fig. 7.3 A sample circuit

In the above circuit, the three voltage sources have been shown on separate branches and on separate loops. However, multiple voltage sources could be on a single branch. For the above circuit, consider Loop1. Loop1 encounters the voltage source V1 while moving from the negative terminal to the positive terminal, and therefore, the voltage source appears as +V1 in Loop1. The first row of matrix B will therefore be the vector

$$\begin{bmatrix} 1 & 0 & 0 \end{bmatrix} \tag{7.9}$$

In the case of Loop2, the voltage source V2 is encountered while moving from positive terminal to negative terminal and therefore appears as -V2. The second row of matrix B will be

$$\begin{bmatrix} 0 & -1 & 0 \end{bmatrix} \tag{7.10}$$

Since Loop3 has no voltage source, the third row will be [0 0 0]. The fourth row of matrix B will be [0 0 -1]. In this manner, matrix B maps the voltage sources in the circuits in the vector **u** to the loop equations. The matrix B is

$$\mathbf{B} = \begin{bmatrix} 1 & 0 & 0 \\ 0 & -1 & 0 \\ 0 & 0 & 0 \\ 0 & 0 & -1 \end{bmatrix} \tag{7.11}$$

With this background, the algorithm to generate the matrices **A**, **E**, and **B** will be described.

1. Each loop corresponds to a row in the matrices **A**, **E**, and **B**. Therefore, if a circuit has N independent loops, matrix **A** and **E** will be of sizes NxN while matrix **B** will be of size NxM where M is the number of voltage sources in the circuit.
2. Consider loop j and therefore row j of the matrices. As said before, the diagonal element is determined differently as opposed to the off-diagonal elements. Thus,

the algorithm has two separate cases for element (j, j) of matrices **A** and **E** and the other elements (j, k) with k not equal to j.

3. For the diagonal element (j, j), iterate through all the branches in row j (Loop j) of LoopMap. If a branch m exists which means if element (j, m) of LoopMap is FW or RV, the total resistance of the branch is added to element (j, j) of matrix **A** and the total inductance of the branch is added to element (j, j) of matrix **E**. As discussed before, there is no concept of positive or negative signs for the diagonal elements of **A** and **E**. These diagonal elements are always positive. If the branch m has a voltage source of index k, check the direction in which the voltage source is encountered by the loop. If it is from negative terminal to positive terminal, a +1 is added to the element (j, k) of matrix **B**. If the direction of the loop is from positive to negative terminal, a -1 is added to the element (j, k) of matrix **B**.
4. For the off-diagonal elements in row j of the matrices. Consider the interaction between loop j and another loop l. Check which branches are in common between the loop j and the loop l. If the direction of the branch is the same with respect to loop j and loop l, the two loops are assisting each other with respect to this branch. The sum total of the resistance of the branch is added to elements (j, l) and (l, j) of matrix **A** and the sum total of the inductance of the branch is added to the elements (j, l) and (l, j) of matrix **E**. On the other hand, if the direction of the branch is the same as the direction of one of the loops but is in a direction reverse to the other loop, the two loops are opposing each other with respect to this branch. The sum total of the resistance of the branch is subtracted from the elements (j, l) and (l, j) of matrix **A** and the sum total of the inductance of the branch is subtracted from the elements (j, l) and (l, j) of matrix **E**.

7.3 Solving the Matrix Equation

Following the discussion on how to obtain the matrices for the loop equations, this section will describe how they are solved. The matrix equation that describes the loop equations of a circuit are written in a generic form as follows:

$$\mathbf{E}\frac{d\mathbf{x}}{dt} + \mathbf{A}\mathbf{x} = \mathbf{B}\mathbf{u} \tag{7.12}$$

The vector **x** contains the loop equations while the vector **u** contains the voltage sources in the circuit. To quickly describe how the equations are solved, let us replace the above matrix equation with a simple scalar equation

$$e\frac{dx}{dt} + ax = bu \tag{7.13}$$

In the above equation, at time t = 0s, all variables are at zero. Though in many cases, having nonzero initial values for variables could speed up the simulation, this circuit simulator starts the system at zero. With a simulation time step of dt, the next simulation instant will be t = dt. At this instant, the values of the voltage sources in vector u are updated and the following equation is obtained:

$$dx(t = dt) = \frac{dt}{e} * (-ax(t = 0) + bu(t = dt)) \tag{7.14}$$

With the value of dx obtained at time t=dt, the loop current at time t=dt can be updated as

$$x(t = dt) = x(t = 0) + dx(t = dt) \tag{7.15}$$

In general,

$$dx(t = t_{n+1}) = \frac{dt}{e} * (-ax(t = t_n) + bu(t = t_{n+1})) \tag{7.16}$$

$$x(t = t_{n+1}) = x(t = t_n) + dx(t = t_{n+1}) \tag{7.17}$$

The simulation therefore proceeds iteratively in the above manner. The concept is similar for the matrix equation except that certain matrix techniques need to be introduced.

Theoretically, the matrix equation describing the loop equations can be solved in a similar manner as follows:

$$d\mathbf{x} = dt * \mathbf{E}^{-1} * (\mathbf{A}\mathbf{x} + \mathbf{B}\mathbf{u}) \tag{7.18}$$

However, except for small circuits resulting in matrices of a fairly small size, calculating the inverse of a matrix is a fairly unstable operation. As a circuit becomes larger and has branches that have small inductances, the matrix **E** will be close to singular. Furthermore, if you take into consideration the fact that all the loop equations are not differential equations, the matrix **E** will be singular. As a matter of fact, for both the examples given in this chapter, the matrix **E** is singular. In order to solve the above matrix equation, we borrow a concept from linear algebra that is very often used in solving a set of linear equations simultaneously. In the case of a matrix equation as the one above, any elementary row operations performed on the matrices on both sides of the equation will not change the solution set. An elementary row operation can be any of the following:

1. Multiplying a row of all the matrices by a nonzero constant.
2. Adding a row to another row.
3. Interchanging two rows.

To ensure that the solution does not change, when one of the above operations is performed, it must be performed on all the matrices in the equations. The next step will be to describe the objective of these row operations.

In the matrix equation, if the matrix **E** has one of the following structures, it is called a triangular matrix.

$$\begin{bmatrix} e_{11} & \times & \dots & \times \\ 0 & e_{22} & \dots & \times \\ \vdots & \vdots & \ddots & \vdots \\ 0 & 0 & \dots & e_{NN} \end{bmatrix} \qquad \begin{bmatrix} e_{11} & 0 & \dots & 0 \\ \times & e_{22} & \dots & 0 \\ \vdots & \vdots & \ddots & \vdots \\ \times & \times & \dots & e_{NN} \end{bmatrix} \tag{7.19}$$

The matrix on the left is called an upper triangular matrix while the one on the right is called a lower triangular matrix. The diagonal elements as well as the elements marked as "x" are elements that can be nonzero or zero. However, for the matrix on the left, all the elements in the lower half of the matrix below the diagonal must be zero and for the matrix on the right, all elements in the upper half of the matrix above the diagonal must be zero. Let us consider the upper triangular matrix on left in detail as this is the matrix relevant to the circuit simulator, though exactly the same results would have been obtained if the lower triangular matrix had been chosen instead. For the upper triangular matrix, the matrix equations can be expanded as below:

$$\begin{aligned} e_{11}\frac{dx_1}{dt} + e_{12}\frac{dx2}{dt} + \dots + e_{1N}\frac{dx_N}{dt} &= A_1x + B_1u \\ e_{22}\frac{dx_2}{dt} + \dots + e_{2N}\frac{dx_N}{dt} &= A_2x + B_2u \\ \vdots &= \vdots + \vdots \\ e_{NN}\frac{dx_N}{dt} &= A_Nx + B_Nu \end{aligned} \tag{7.20}$$

where

$$A = \begin{bmatrix} A_1 \\ A_2 \\ \vdots \\ A_N \end{bmatrix} \quad \text{and} \quad B = \begin{bmatrix} B_1 \\ B_2 \\ \vdots \\ B_N \end{bmatrix} \tag{7.21}$$

The matrices **A** and **B** are written as a collection of row vectors as their structure is not of importance. As can be seen from the left-hand side of the equation, since matrix **E** is in upper triangular form, it is in a form such that it can be solved backwards in a fairly convenient manner. Starting from the last row, the derivative of x_N can be calculated to be substituted in the second last row to calculate the derivative of x_{N-1}. In this manner which can now be seen similar to the solving of the simple scalar differential equation, all the derivatives of **x** on the left-hand side can be calculated at time t=tk+1 while the values of **x** on the right-hand side are at time t=tk with the values of the input **u** also at time t=tk+1.

After having explained the advantages of converting matrix **E** to an upper triangular form, we will now describe the algorithm to do so. As stated before, the elementary row operations performed on matrix **E** to achieve this conversion have to also be performed on matrix **A** and matrix **B** in the matrix equation for the solution to be unchanged. Let us begin by considering the following generic matrix **E**:

$$\mathbf{E} = \begin{bmatrix} e_{11} & e_{12} & \dots & e_{1N} \\ e_{21} & e_{22} & \dots & e_{2N} \\ \vdots & \vdots & \vdots & \vdots \\ e_{N1} & e_{N2} & \dots & e_{NN} \end{bmatrix} \tag{7.22}$$

To convert this matrix to an upper triangular form, we consider each diagonal element e11 to eNN. The objective is to make all the elements below a diagonal element to be zero using row operations. For example, to make e21 zero, the operation Row2 = Row2 - e21*Row1/e11 can be performed. The same for the remaining rows Row3 to RowN. There is however the possibility that e11 is zero but some of the other elements below e11, i.e., e21 to eN1 are not zero. When e11 is zero, it is impossible through row operations to make any of the elements below it zero through row operations. Therefore, the only solution in this case is to exchange row 1 which contains e11 with any one of the rows that contain a nonzero element in the first column. After this, the row operations can be performed to make all the elements below the new e11 zero. The result of performing row operations from row 2 to row N will be the matrix:

$$\mathbf{E} = \begin{bmatrix} e_{11} & e_{12} & \dots & e_{1N} \\ 0 & e'_{22} & \dots & e'_{2N} \\ \vdots & \vdots & \vdots & \vdots \\ 0 & e'_{N2} & \dots & e'_{NN} \end{bmatrix} \tag{7.23}$$

Row 1 will not change unless a row interchange has been performed in case element e11 is zero. However, the remaining rows have changed. By repeating the above sequence of row operations on row 2 to row N-1 with respect to the remaining diagonal elements, the matrix can be reduced to an upper triangular form.

The last step of this section is a short discussion on the numerical method used in solving the differential equations. Though the method used can have a profound impact on the solution in terms of accuracy, speed of arriving at a solution, and also stability of the solver, the topic of solving differential equations is vast and is largely out of the scope of this book. The focus of this book is the understanding of how circuits can be represented and simulated. In the circuit simulator, the numerical method can be found in the function mat_ode in the file solver.py. Three solvers have been tested and tried and a brief description is as follows. For a more detailed description, the reader is encouraged to read separately on the topic.

1. Trapezoidal rule: This is fairly basic solver which employs the technique shown in the description at the beginning of this section. It involves the calculation of one slope dx/dt which is then used to update the loop equations x. The solver is lightweight but as the complexity of the circuit increases, it requires smaller time steps to be able to simulate a circuit without becoming unstable.
2. Runge–Kutta Fourth-Order Method: This is the solver which is used by the circuit solver. It is more advanced compared to the trapezoidal rule as in with this solver, it calculates four slopes of the differential equation and uses the weighted average

to update the solution. It is heavier in terms of computing burden but provides fairly stable performance for reasonable values of simulation time step.

3. Runge–Kutta Fifth-Order Method: This is an advanced version of the previous fourth-order method, and in this solver, five slopes of the differential equation are calculated. The solver is quite heavy in terms of computational burden and though has been tested briefly is not normally used as the simulations have significantly longer execution times.

The solution of the matrix equation will provide the values of the loop currents. However, these loop currents are for the purpose of circuit analysis and the currents whose values are needed are the branch currents. The next section will describe how the branch currents can be calculated from the loop currents.

7.4 Mapping Branch Currents and Loop Currents

As said in the concluding part of the previous section, the loop currents are for the purpose of analysis but it is the branch currents that are of prime importance. For example, an Ammeter in a branch needs to have the current through it calculated. Or a Diode in a branch must have the current in the branch calculated to be able to determine when it will turn off. This section will describe the method by which branch currents are mapped to loop currents. Furthermore, as will be shown later in the book when we examine the nonlinearity in circuits, there will be a need to determine loop currents at a particular instant from the branch currents. A more detailed description of this need will be provided in the subsequent chapters. But this section will describe how the loop currents can be calculated from the branch currents of the circuit.

Let us consider the circuit in the previous section which is shown in Fig. 7.4. The LoopMap for the circuit in Fig. 7.4 is shown in Fig. 7.5. The LoopMap has been

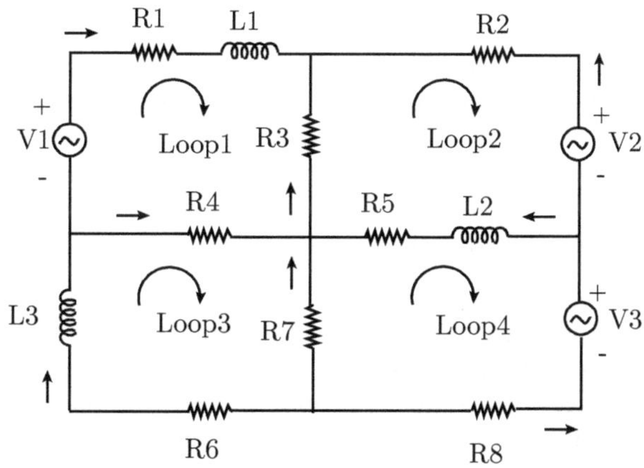

Fig. 7.4 Showing relationship between loop currents and branch currents

	{V1,R1,L1}	{V2,R2}	R3	R4	{R5,L2}	{R6,L3}	R7	{V3,R8}
Loop1	FW		RV	RV				
Loop2		RV	FW		FW			
Loop3				FW		FW	RV	
Loop4					RV		FW	RV

Fig. 7.5 LoopMap

used so far to generate the system matrices to perform loop analysis. However, the LoopMap essentially provides information of the branches in each loop and their direction with respect to the direction of the loop. Therefore, this information can be used to calculate branch currents from loop currents. For example, in LoopMap above, the branch {V1, R1, L1} appears only in Loop1 in which its direction is the same as the direction of Loop1. The current through the branch is thus the same as the loop current. On the other hand, branch {R3} appears in Loop1 and Loop2 with its direction being the same as the direction of Loop2 and opposite to the direction of Loop1. Therefore, current in branch R3 will be i2-i1. In general, the algorithm for calculating the branch currents from the loop currents is as follows:

1. Iterate through all the branches in the circuit.
2. For any branch k, initialize its current to zero.
3. Iterate through all the loops while checking if the branch exists in that loop. If the branch is in the same direction (FW) as that of the loop, the loop current is added to the branch current. If the branch direction is opposite (RV) to the direction of the loop, the loop current is subtracted from the branch current.

Now to describe the reverse process—the calculation of the loop currents from the branch currents. This process also uses LoopMap but can be divided into two parts. In almost every LoopMap, there will be two cases related to the presence of branches in loops. First, there will be those branches that are present only in a single loop like branch {V1, R1, L1} for example. Second, there will be those branches that will be present in two or more loops like branch {R3} for example. For the branches that are found only in one loop, there exists a one-to-one relationship between the branch current and the loop currents. Therefore, current in branch {V1, R1, L1} can be calculated from current in Loop1 and vice versa. This can be formulated with the following algorithm:

1. Iterate through all the loops.
2. For a particular loop k, iterate through all the branches. If a branch exists in the loop, check if it exists in other loops. If it does not exist in other loops, it means there is a one-to-one relationship between this branch and the loop.
3. Add the loop and the branch to a pair to mark this one-to-one relationship.

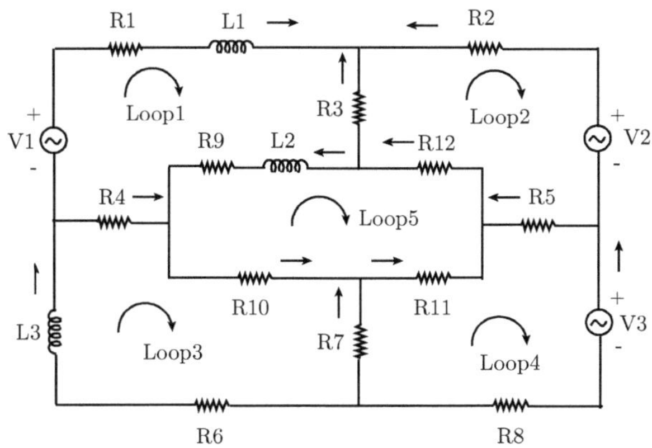

Fig. 7.6 A loop that has no unique branch

	{V1,R1,L1}	{V2,R2}	R3	R4	R5	{R6,L3}	R7	{V3,R8}	{R9,L2}	R10	R11	R12
Loop1	FW		RV	RV					FW			
Loop2		RV	FW		FW							FW
Loop3				FW		FW	RV			FW		
Loop4					RV		FW	RV			FW	
Loop5									RV	RV	RV	RV

Fig. 7.7 LoopMap for the circuit

The reason for the above process is that in any circuit, many branches will be present only in one loop. Therefore, the above one-to-one relationships will be applicable for many branches if not a vast majority. As compared to the procedure described below for the loops that only have branches which are in two or more loops, the above process has no computation but only checking for the presence or absence of branches. This reduces the overall computational burden.

Now to describe the process of calculating the currents of those loops that only have branches in two or more loops. Since there is no such loop in the circuit of Fig. 7.4, consider the circuit in Fig. 7.6. By mere visual inspection, the inner loop 5 has no branch that exists only in that loop which can be confirmed by the LoopMap in Fig. 7.7. From the above LoopMap, the current in Loop5 can be calculated from any of the four equations:

$$
\begin{aligned}
i_{loop1} - i_{loop5} &= i_{\{R9,L2\}} \\
i_{loop3} - i_{loop5} &= i_{\{R10\}} \\
i_{loop4} - i_{loop5} &= i_{\{R11\}} \\
i_{loop2} - i_{loop5} &= i_{\{R12\}}
\end{aligned}
\tag{7.24}
$$

In the above case, except for iloop5, all the other loop currents have been determined from branch currents as they have one-to-one relationships with branches. It may so happen that there are multiple loops that need to be calculated as above. In that case, these loop currents will need to be calculated by solving the equations simultaneously as matrix equations in a manner similar to the method described above except that the matrix equations will be in the form of:

$$\mathbf{A}\mathbf{x} = \mathbf{b} \tag{7.25}$$

In the above equation, **x** will contain all the currents in the loops that do not have a one-to-one relationship with a branch, while **b** will contain the branch currents and the loop currents that can be determined from one-to-one relationships with branches.

This section concludes the basic concept of simulating the circuit using loop analysis with which both loop currents and the currents in every branch of the circuit can be calculated. The next section will describe how this concept may be sufficient for extremely simple linear circuits but nonlinearity and time constants will need additions to the simulator.

7.5 Effects of Time Constants on Loop Analysis

The previous section described how loop analysis is performed on a circuit by generating the system matrices from the LoopMap. The previous process would work with acceptable accuracy and under reasonable time durations if the parameters of the circuit are not too vastly different. In this section, we now describe how the parameters of the circuit determine the time constants of the circuit and the impact they have on simulations. Consider the very simple circuit in Fig. 7.8. The voltage source V is a sinusoidal voltage of frequency 60 Hz and RMS voltage magnitude 120 V. Let us start with a resistance of R=50 ohm and L=0.001 H. We shall choose a simulation time step of 10 μs and a time duration of 0.1 s. The differential equation for the current in the above circuit will be:

$$V - Ri - L\frac{di}{dt} = 0 \tag{7.26}$$

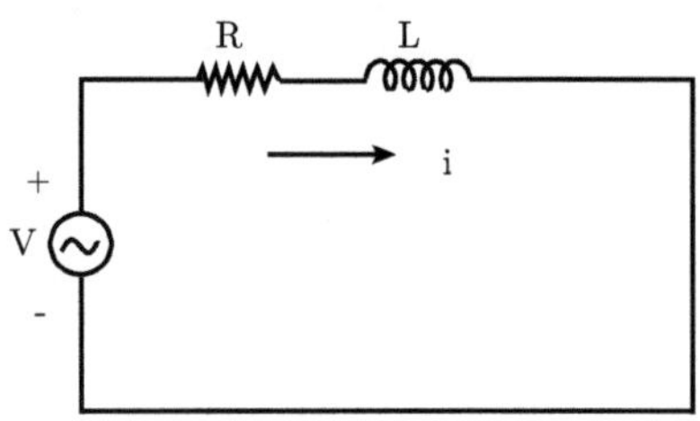

Fig. 7.8 A simple R-L circuit

Therefore, the current update equation at time $t = t_{k+1}$ will be

$$i(t = t_{k+1}) = i(t = t_k) + \frac{V(t = t_{k+1}) - Ri(t = t_k)}{L} dt \qquad (7.27)$$

The time constant of the above circuit will be from basic network principles L/R. In the update equation above, the time constant will therefore play a role as the ratio L/R can be seen to appear if the equation is expanded. To describe the role of the time constant, we will now look at how the current generated by the simulation changes when the resistance is increased, resulting in decreasing time constants. Figure 7.9 shows the voltage and the current waveform together. As it is the current waveform that we are mainly interested in, let us consider that in Fig. 7.10. The above simulation can be seen to be stable. The time constant of the circuit is 0.001/50 = 200 μs. Since the simulation time step is 10 μs, it should be small enough to simulate the circuit with a time constant of 200 μs. Now, let us increase the resistance R to 200 ohms. The simulation is still stable as seen from Fig. 7.11. However, the time constant of the circuit is now 0.001/200 = 20 μs. The next simulation result will show that this is a borderline case. Let us increase the resistance further to 200.1 ohms.

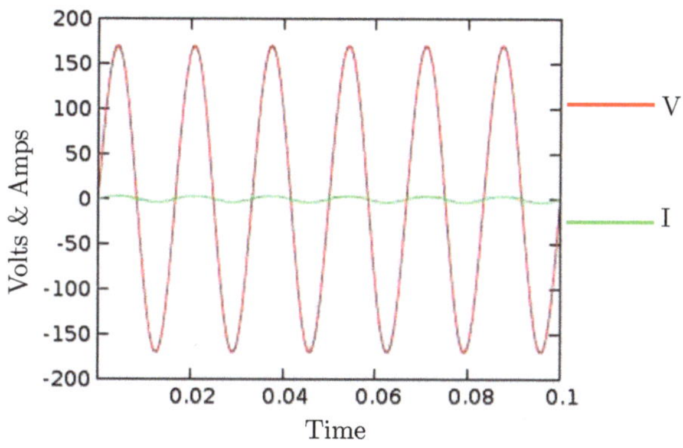

Fig. 7.9 Result of a stable simulation

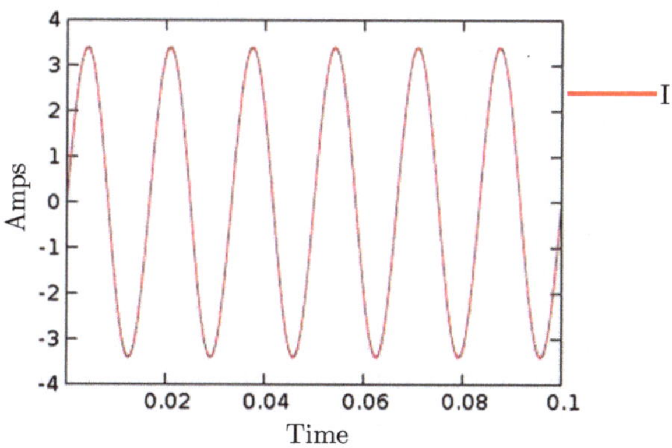

Fig. 7.10 Current waveform in the stable simulation

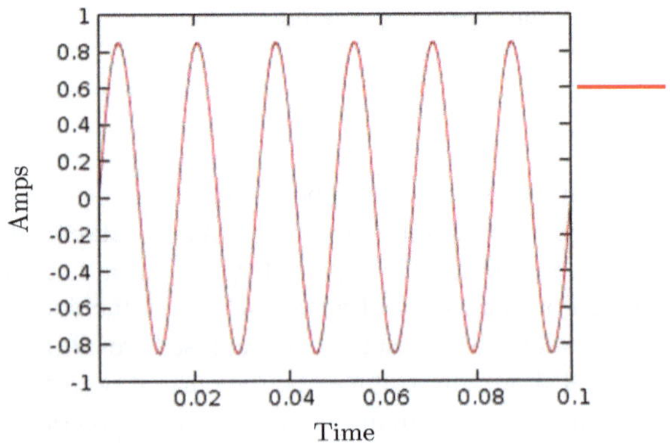

Fig. 7.11 Current waveform in the marginally stable simulation

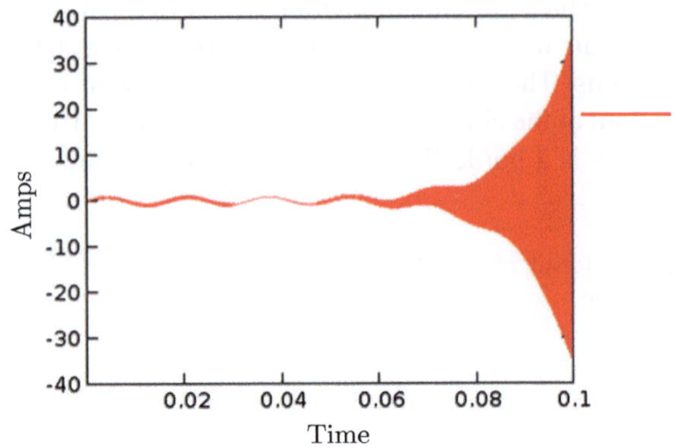

Fig. 7.12 Current waveform in the unstable simulation

The result in Fig. 7.12 shows the simulation becoming unstable. The circuit taken for the example is fairly simple, and therefore, it is possible to examine the time constant of the circuit and compare it with the simulation time step. In this example, the reason for instability is that the simulation is not fast enough to capture the dynamics of the circuit—10 μs versus 19.9 μs. A simple way to solve this problem would be to decrease the simulation time step. Figure 7.13 shows the simulation result with a time step of 5 μs. However, this solution cannot be used for arbitrary values of circuit parameters as this might mean the simulation time step would have to be unrealistically small. Usually for power electronics circuits, a simulation time step of 1 μs should be sufficient while for high-frequency converters, a time step of 100 ns is acceptable. However, if time steps smaller that these are needed, either the application should be a special one or the circuit parameters need to be modified to improve the stability of the simulation.

Differential equations with extremely low time constants are called stiff equations. In a large circuit, one or more branches could be such that they have a low time constant because the L/R ratio of the branch is extremely low. These branches result in loop equations that end up having extremely low time constants and therefore

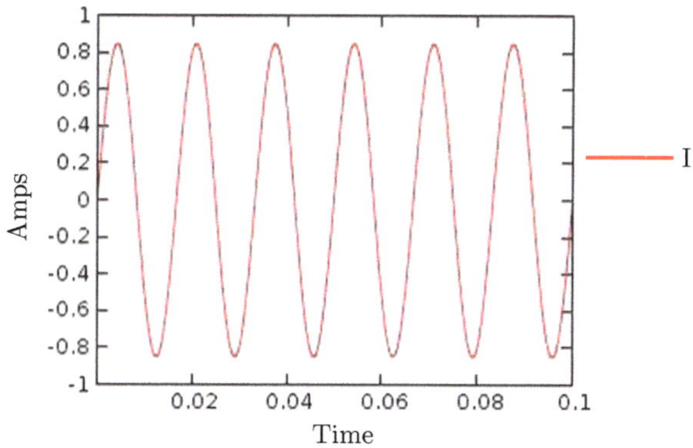

Fig. 7.13 Current waveform when time step is decreased

are called stiff branches. There are usually two reasons for stiff branches. First, some branches have only a parasitic inductance which may be in the order of a few microhenries. For example, a load modeled by a resistor of 20 ohm and a parasitic inductance of 10 microhenry. Second, a branch has a very large resistor such that it causes the branch to be an open circuit or to draw a negligible current. For example, a Voltmeter is modeled as a large resistor so that it draws no more than a microampere of current for the maximum possible voltage. This results in the Voltmeter having a resistance in the order of megaohms. Another example would be a Diode which becomes a high resistance when it turns off so that it draws negligible current at the maximum rated voltage. In both these cases, any loop that contains a Voltmeter or a Diode in the off state will have an extremely low time constant for any practical value of inductance. Since measuring the voltage in a circuit is something any user would like as much as possible and since this simulator is targeted specifically toward power electronic circuits, it is essential that the simulator be able to deal with stiff branches without becoming unstable.

After a description of the problem of instability caused by stiff branches and loops, the next section will describe the solution employed in the circuit simulator.

7.6 Effect of Stiff Loops

In loop analysis, if a circuit is represented by a set of linearly independent loop equations, each branch in the circuit will appear in at least one equation. A branch may be present in multiple equations and in an extreme case in a majority of the equations. The first step in dealing with stiff branches is to limit the presence of stiff branches to the minimum possible number of equations. To illustrate this concept, consider the circuit in Fig. 7.14. The circuit contains a Voltmeter Vm. This Voltmeter is an extremely large resistance such that it will draw a negligible current at its maximum specified voltage. This Voltmeter is therefore a stiff branch in the circuit. Let us assume that all other branches are not stiff and have time constants that are

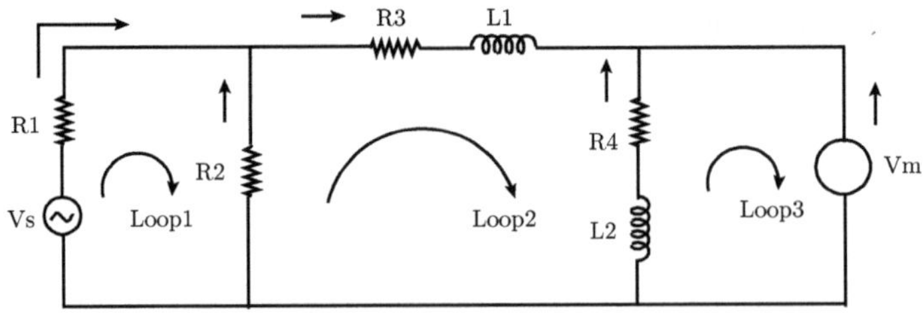

Fig. 7.14 A sample circuit with one stiff branch

reasonably large. The Voltmeter appears in only Loop3. This Loop L3 will have a time constant of L2/(R4+Rvm) which is negligible and will need an extremely low simulation time step to be able to produce a stable simulation. To avoid this and continue with the specified simulation time step, the only solution is to ignore the inductance L2 in Loop 3 and solve it as a static equation instead of a differential equation. By doing so, instability is ruled out but at the cost of accuracy. The Voltmeter Vm measures the voltage across the branch containing R4 and L2. This implies the voltage drop across R4 and L2. By ignoring L2, the voltage measured by Vm will be either slightly inaccurate or blatantly wrong depending on how large the drop across L2 is compared to R4. This problem is more complex and will be described later. However, first we address the problem of limiting a stiff branch to the minimum number of loops.

Consider the circuit in Fig. 7.15 which is merely the loop equations for the above circuit rewritten in a different manner. By mere visual inspection, it can be seen that the loops are independent and would adequately represent the circuit. The loop search algorithm is a random search algorithm that starts with different branches and looks for closed paths. Therefore, depending on how a circuit is drawn and the position of the branches, a circuit drawn in different ways can lead to very different set of loops. Therefore, the set of loops in Fig. 7.15 is quite possible. The problem now is that every loop contains the Voltmeter Vm and therefore is stiff. If the previous solution of ignoring all the inductances in the loops is employed, the circuit will have no dynamics, and with all the loops having extremely large resistances, the loop currents and the resulting branch currents will be extremely small—all of these

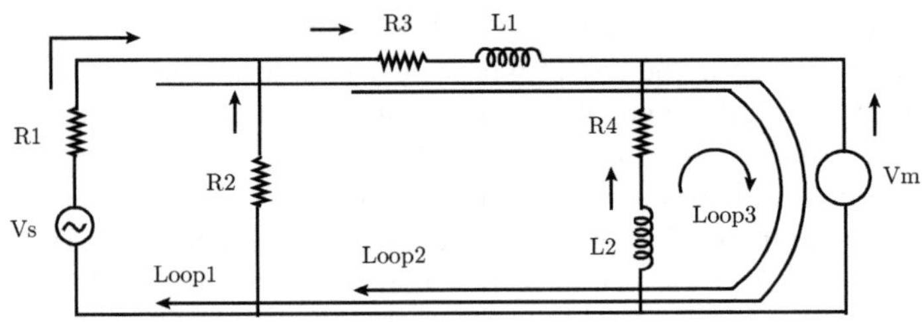

Fig. 7.15 All loops written with respect to stiff branch

Fig. 7.16 LoopMap for above set of stiff loops

	{Vs,R1}	R2	{R3,L1}	R4L2	Vm
Loop1	FW		FW		RV
Loop2		FW	FW		RV
Loop3				FW	RV

Fig. 7.17 Modified LoopMap

	{Vs,R1}	R2	{R3,L1}	{R4,L2}	Vm
Loop1	FW	RV			
Loop2		FW	FW	RV	
Loop3				FW	RV

are completely wrong results. Therefore, the first problem that needs to be solved is that the Voltmeter Vm cannot appear in all loops. To achieve this, we return to the LoopMap and introduce the concept of loop manipulations.

The LoopMap for the circuit in Fig. 7.15 can be written as shown in Fig. 7.16. For the above circuit, elementary operations on the loop equations could be performed to eliminate Vm from some of the equations. For example, the operations:

```
Row1 = Row1 - Row3
Row2 = Row2 - Row3
```

These operations would result in the LoopMap shown in Fig. 7.17. An examination of the above LoopMap will show that the loops are exactly the loops depicted in the circuit of Fig. 7.14 where Vm appears only in Loop 3. Therefore, using row operations on the LoopMap, we have managed to isolate Vm to a single loop where the assumption of ignoring L2 and solving Loop 3 as a static equation can be employed. Now that this concept has been illustrated with an example, a detailed description of this process will be provided.

7.7 Loop Manipulations

For a circuit having B branches and N nodes to be completely represented using loops, the number of independent loops will be B-N+1. Each branch will have to appear in at least one of these equations. If a branch does not appear in even a single equation, the set of loop equations do not accurately represent the circuit as the dynamics of a branch have been neglected. Conversely, if a branch appears in a single loop and the loops remain independent, the loops will continue to accurately represent the branch. Therefore, it is possible to manipulate the loops so as limit the presence of certain branches to a minimum number of loops preferably even a single loop. This is the concept used with stiff branches and has been demonstrated in the previous section where the loops were manipulated so as to limit the Voltmeter Vm to a single loop.

Before we present the concept of loop manipulations, it is necessary to distinguish between loop manipulations and elementary operations on equations. Let us consider a simple case of N equations with N variables written in the following matrix form:

$$\mathbf{Ax} = \mathbf{B} \tag{7.28}$$

Here **x** is the vector of the N variables, and **A** is an NxN matrix while **B** is a vector of size N. An elementary row operation performed on the above set of equations will not change the solution set. Therefore, if the above equations are independent, this would imply that the matrix **A** is invertible. Similarly, if the equations were not independent, the matrix **A** would not be invertible, and through a succession of row operations, it will be possible to show that one or more rows of the matrix **A** will be zero. However, in any case, row operations performed on the above set of equations will not change the solution.

On the contrary, the same concept cannot be directly applied to loops. A collection of loops in a LoopMap cannot be treated in the same manner as equations particularly when it comes to applying row operations. A loop is a set of branches that forms a closed path in an electrical circuit. If a row operation is performed on two loops, for example, a loop subtracted from another loop, the resultant set of branches need not form a loop. This is because these branches may not form a closed path or may violate any of the other conditions of a loop such as a repetition of nodes. Consider the following example of loops in Fig. 7.18. In this circuit, the loop manipulation Loop2 - Loop1 is performed. But from visual inspection, it can be seen that the result is not a single closed loop but two loops Loop3' and Loop3" and therefore unacceptable as a result. Several other examples are possible. For this reason, it is necessary after a loop manipulation to check if the resultant set of branches forms a loop according to the guidelines laid out for a loop:

1. It should be a closed path—should have one node as starting and ending node.
2. Should not have a repetition of any other node as this results in subloops.

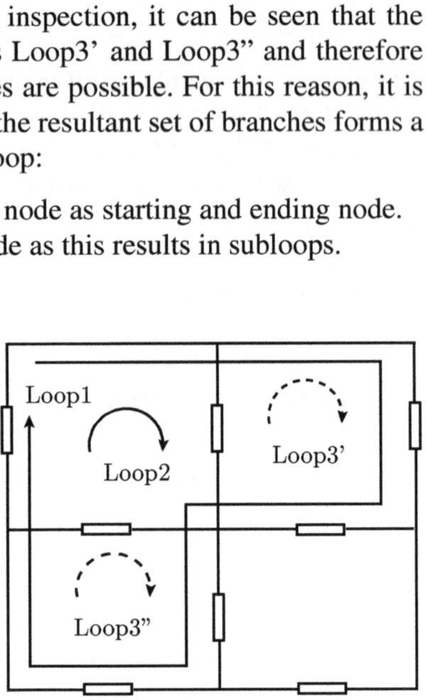

Fig. 7.18 Incompatible loops

3. Each node should appear in exactly two branches—more would again mean subloops as this means there is a fork instead of a single closed path while less than two means there is a break in a loop.

The logic used for loop manipulations is fairly intuitive. To begin with, two loops can only be manipulated when there is at least one branch in common between the loops because otherwise the two loops are completely separate. If there are N branches in the circuit and Loop k and Loop j are being manipulated, the branch m of the resultant loop will be decided as follows:

1. If branch m exists in both Loop k and Loop j, Loop k - Loop j will result in the branch being deleted if (Branchm)Loopk = "FW" and (Branchm)Loopj = "FW" or (Branchm)Loopk = "RV" and (Branchm)Loopj = "RV", i.e., the branch is in the same direction in both loops. On the other hand, Loop k + Loop j will result in the branch being deleted if (Branchm)Loopk = "FW" and (Branchm)Loopj = "RV" or (Branchm)Loopk = "RV" and (Branchm)Loopj = "FW", i.e., the loops are opposing each other with respect to that branch.
2. If a branch exists in only one loop, for example, in Loop k only, Loop k - Loop j will result in the same branch that Loop k has. Which means, if (Branchm)Loopk = "FW" and (Branchm)Loopj = "-", the resultant loop will have (Branchm)result = "FW" while if (Branchm)Loopk = "RV" and (Branchm)Loopj = "-", the resultant loop will have (Branchm)result = "RV".

Let us now examine the loop manipulation performed between Loop 1 and Loop 3 for the circuit in Fig. 7.15 shown in the LoopMap of Fig. 7.19. The objective of the loop manipulation is to eliminate the branch Vm from Loop 1. Since Vm appears in the same direction "RV" in Loop 1 and Loop 3, the manipulation performed is Loop 1 - Loop 3. In the case of branches {Vs, R1} and {R3, L1}, Loop 1 has these branches as "FW" while they do not exist in Loop 3. The result in both cases is therefore a "FW". In the case of branch {R4, L2}, the branch does not exist in Loop 1 but is a "FW" in Loop 3. The result is therefore a "RV". Since Vm exists in both loops as "RV", the result is that it is eliminated as expected. The remaining branches do not exist in either of the loops and therefore will not exist in the result either.

As shown in the above example, the branch with Voltmeter Vm is eliminated from Loop 1 and Loop 2. The circuit was a simple one but for a complex circuit with several stiff branches, an algorithm will need to be developed. The first step of this algorithm is to mark the branches that are stiff on the LoopMap. If a branch is

Fig. 7.19 Loop manipulation in a LoopMap

	{Vs,R1}	{R3,L1}	{R4,L2}	Vm	
Loop1	FW	FW		RV	
Loop3			FW	RV	
Result	FW	FW	RV		

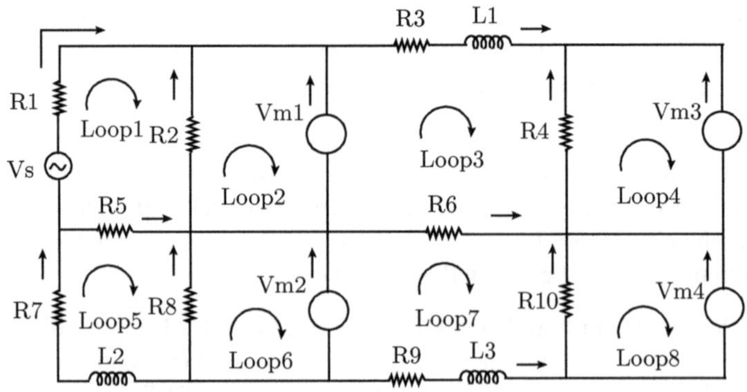

Fig. 7.20 A circuit with several loops

	{Vs,R1}	R2	Vm1	{R3,L1}	R4	Vm1	R5	R6	{R7,L2}	R8	Vm3	{R9,L3}	R10	Vm4
Loop1	FW	RV					RV							
Loop2		FW	SR											
Loop3			SF	FW	RV			RV						
Loop4					FW	SR								
Loop5									FW	RV				
Loop6										FW	SR			
Loop7								FW			SF	RV	RV	
Loop8													FW	SR

Fig. 7.21 LoopMap

stiff and is in the same direction as the direction of the loop, it will be marked as "SF" while if the direction of the branch is opposite to that of the loop, it will be marked as "SR". Let us consider a more complicated circuit in Fig. 7.20 with four stiff branches as Voltmeters. The LoopMap for the circuit with the loops is shown in Fig. 7.21. As can be seen from the LoopMap, the Voltmeters Vm2 and Vm4 appear only in Loop 4 and Loop 8, respectively, and do not need to eliminate from any other loops. However, the Voltmeters Vm1 and Vm3 appear in multiple loops—Vm1 in Loop 2 and Loop 3 while Vm3 in Loop 6 and Loop 7.

The first part of the process of loop manipulations is to bring all the loops containing stiff branches to the top of the LoopMap. This will help in solving the loop equations as the loops are now segregated as those stiff loops whose time constants have been deliberately ignored versus the remaining loops. Let us now examine only the first six stiff loops of this LoopMap as shown in Fig. 7.22. There are four stiff

	{Vs,R1}	R2	Vm1	{R3,L1}	R4	Vm1	R5	R6	{R7,L2}	R8	Vm3	{R9,L3}	R10	Vm4
Loop1		FW	SR											
Loop2			SF	FW	RV			RV						
Loop3					FW	SR								
Loop4										FW	SR			
Loop5								FW			SF	RV	RV	
Loop6													FW	SR

Fig. 7.22 Manipulated LoopMap

branches in the circuit, the objective is to restrict these stiff branches to the minimum number of loops. Since there are four stiff branches, the minimum number of loops needed to contain these branches would also be four as every branch of the circuit must appear in at least one loop. Therefore, in a manner similar to solving loop equations, we try to make this LoopMap diagonal with respect to the stiff branches. The algorithm will be as follows:

1. Start with each stiff loop. When a stiff loop has been found, look for the first stiff branch in that loop. This stiff branch is to be treated as a diagonal element and with row operations, the objective is to eliminate this stiff branch from all other stiff loops.
2. For example, Loop 1 is a stiff loop. The first stiff branch is Vm1. The stiff branch also exists in Loop 2; however, the direction of the branch is against the loop in Loop 1 and in the same direction as Loop 2. Therefore, the row operation performed is Loop2 = Loop2 + Loop1. The loop operation is shown below in Fig. 7.23, and it can be seen that the result is a valid loop and does not have any stiff branches. This result is then assigned to Loop2.
3. The next stiff loop is Loop 3 but the only stiff branch is Vm2. However, Vm2 does not appear in any other loops and therefore there is no need to perform any loop manipulation.
4. For Loop 4 and Loop 5, loop manipulations are needed as Vm3 is present in both these loops. These can be performed in the same manner as above.
5. Finally, Loop 6 has only one stiff branch Vm4 and Vm4 does not appear in any other loop and can be left as it is.

Using the technique of loop manipulations, the stiff branches can be limited to four loops with a stiff branch in each loop. However, in the case of circuits with several stiff branches and only a few non-stiff branches, it may so happen that multiple stiff branches appear in a single loop. This is because the number of branches in a circuit are typically more that the number of loops, and therefore if it so happens that the number of stiff branches is greater than the number of loops, it is impossible to have

Fig. 7.23 Process of loop manipulation

	R2	Vm1	{R3,L1}	R4	R6
Loop1	FW	SR			
Loop2		SF	FW	RV	RV
Result	FW		FW	RV	RV

a single stiff branch per loop. In that case, even though multiple stiff branches appear in a single stiff loop and also a stiff branch may appear in multiple stiff loops, the technique helps to draw out the non-stiff loops that do not have a single stiff branch. The differential equation solver can now solve these non-stiff loops. In the later chapters, it will be shown how in some circuits with power electronic converters, non-stiff loops that represent the circuit only appear at a later stage of the simulation.

This covers most of the aspect of loop analysis. At this point, we would like to describe how loop analysis alone is not sufficient to simulate a nonlinear circuit. The next section will describe this aspect and will lay the foundation for the next chapter which will describe how nodal analysis is used in the simulator.

7.8 Limitation of Loop Analysis

Loop analysis is fairly accurate when all the branches and subsequent loops in the circuit are non-stiff. However, with stiff branches in a circuit, the resulting stiff loops have to be approximated as shown in the previous section. In this section, we shall describe this problem in detail. To begin with, a stiff branch could have a large resistance for several reasons—it could be a Voltmeter, it could be a nonlinear element (a Diode) in the off state, or it could be a large resistance the user wants to use to break a branch for testing purposes. Since, the main characteristic of a stiff branch is that it is purely resistive, the current and voltage in a stiff branch will have a simple V/I=R relationship. As will be shown below, it becomes more of a problem of determining the voltage across the stiff branch rather than determining the current in a stiff branch.

Let us reconsider the circuit in Fig. 7.24 that contained the single stiff branch as Vm. Loop 3 as mentioned before will have an extremely small time constant L2/(R4+Rvm). This is neglected by ignoring inductance L2. This might seem a reasonable approximation, but if we consider the possibility that the voltage drop across the inductor L2 may be as high as that of the resistor R4, ignoring the inductance could result in a blatantly wrong result. This is because the current in Loop 3 will be the same as the current through the branch Vm, and therefore, voltage measured by the Voltmeter would be iloop3*Rvm. Therefore, the Voltmeter Vm does not measure the voltage drop across the branch but only the voltage drop across the resistor R4. This problem can be minimized by not completely ignoring the inductance from

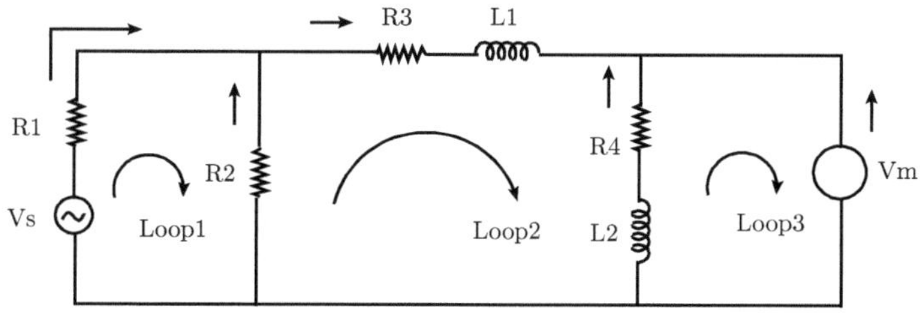

Fig. 7.24 A circuit with a stiff branch

Fig. 7.25 Stiff branches in parallel with non-stiff branches

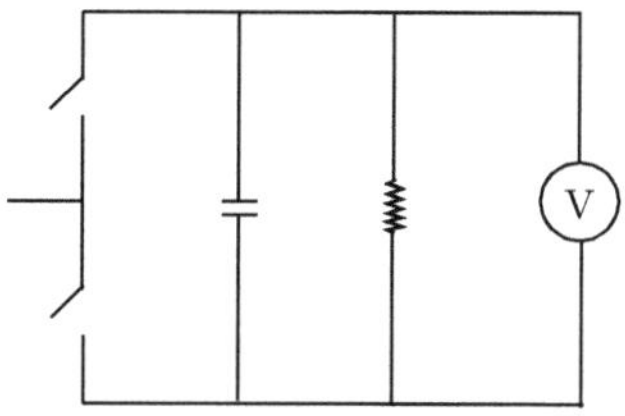

calculations of stiff loop. In the above case, the complete loop equation for Loop 3 can be written as:

$$R4 * i_{loop3} + Rvm * i_{loop3} + L2 * \frac{di_{loop3}}{dt} = R4 * i_{loop2} + L2 * \frac{di_{loop2}}{dt} \tag{7.29}$$

Since Loop 3 is a stiff loop with a very low time constant—L2/(Rvm+R4) being negligible due to the high resistance Rvm of Voltmeter Vm. Therefore, di_{loop3}/dt is ignored resulting in the following approximation:

$$R4 * i_{loop3} + Rvm * i_{loop3} = R4 * i_{loop2} + L2 * \frac{di_{loop2}}{dt} \tag{7.30}$$

From the above equation, the impact of the voltage drop across the inductor is introduced into the calculation of the current in Loop 3 due to the term $L2 * di_{loop2}/dt$. This results in a fairly accurate Voltmeter reading.

There is, however, a far more deep-rooted problem with stiff branches. This comes from the fact that stiff branches are isolated to the minimum possible number of loops to reduce the impact on the remainder of the loop equations. Though this is technically correct, in a numerical simulation, it is a challenge with respect to stability. Consider the segment of a circuit shown in Fig. 7.25. The two Switches shown on the left are a part of a larger circuit with power electronic converters. When a Voltmeter is connected across parallel branches in such a manner, the voltage across these parallel branches are always the same. However, one of the branches has a dominant effect on the voltage. For example, if both the Switches are off, the capacitor voltage dominates while if both Switches are on, the capacitor gets short-circuited and even if it has a voltage, the voltage drop across the branches is almost zero. If the technique of loop manipulations above is used and the Voltmeter branch is restricted to a single loop, it will be associated with only one of the branches in the group of branches connected in parallel. Therefore, the cumulative effect of the parallel-connected branches will not be fully captured and in the above case, even more so as one of the branches is nonlinear. Simulations of circuits similar to the one above showed jitter in the Voltmeter reading when using loop analysis, and the only solution was to decrease the simulation time step to decrease errors in the loop analysis.

Another solution is to examine a technique which calculates the voltages across the branches or in other words the voltages at the nodes. This implies performing a nodal analysis as well. The concept of nodal analysis is explained in the next chapter. As will described, nodal analysis is not only used to deal with stiff branches but also with nonlinear circuits.

7.9 Conclusions

As stated previously, LoopMap represents the loops in a circuit as a collection of branches. This convenient representation allows the circuit simulator to be able to perform a host of computations. To begin with, while generating the loop equations, the LoopMap enables the simulator to determine the total resistance, total inductance, and the total voltage in the closed path that the loop represents. Besides, LoopMap allows for quick determination of the interaction between loops—loops that assist each other over a set of branches or loops that oppose each other over a set of branches. This in turn makes it convenient to generate an algorithm to generate the loop equations and express it as a matrix equation. The chapter has described this process in detail with sample circuits.

Though the solution of matrix differential equations is out of the scope of this book, a description of the method of solving the matrix equations generated from LoopMap has been provided. Significantly, the transformation of the matrices of the equation using elementary row operations before solving them has been described. The solution of the matrix equation is the currents in all the loops of LoopMap. However, for necessary post-processing, the currents in all the branches of the circuit need to be determined. Once again, LoopMap provides information of the branches present in every loop and also thereby provides information of the loops that pass through a branch. This allows for computation of the branch currents from the loop currents by taking an algebraic sum of the currents of the loops in which a branch appears. Conversely, it has also been shown how the reverse calculation is possible. LoopMap can be used to determine the currents in the loops from the currents in the branches.

Since this circuit simulator is primarily intended for power electronics applications, it is extremely important that it is able to deal with constantly changing circuits. The chapter introduces the concept of a stiff branch—a branch with an extremely low time constant (L/R) due to either a negligible inductance or a very large resistance. A power electronics device, such as a Diode, in its OFF state, will be a very large resistance of such a value that it will not allow current to pass through it. Therefore, the ability to deal with stiff branches and branches that change their state from being stiff to non-stiff and vice versa is extremely important. The chapter describes through a simple example how stiff circuits can cause instability if not dealt with appropriately. The technique used to deal with stiff branches is a combination of loop manipulations and approximation.

The chapter has shown how loops are generated in an arbitrary manner and quite often can result in all the loops in LoopMap being stiff even though only a few branches are stiff. Through the process of loop manipulations, the loops can be transformed such that the stiff branches are isolated to the minimum possible number of loops. Therefore, a circuit can be effectively segregated into stiff and non-stiff parts with a segment of LoopMap having only branches that are not stiff while the other segment has the stiff branches. This separation then makes it possible for the stiff loops to be approximated as static equations while the non-stiff loops are solved as differential equations.

Though loop manipulations are an effective technique to ensure stability of a simulation in the presence of stiff branches, quite often they are not sufficient in determining the current through stiff branches accurately. This is for the simple reason that stiff loops are isolated and then approximated which results in the currents of stiff loops not being an accurate representation of the currents in the stiff branches. The chapter provides an example where a jitter may be experienced in the measurement of Voltmeters in circuits with parallel branches that contain nonlinear elements. To solve this problem, the circuit simulator separately calculates the voltages at the nodes of the circuit from which the branch currents can be determined. The next chapter will describe this process through nodal analysis which also plays a critical role in determining conduction of nonlinear devices in a circuit.

Chapter 8
Circuit Analysis—Nodal Analysis

Abstract This chapter describes how nodal analysis can be used to determine the currents through stiff branches (that have a very low time constant) in the circuit. With the example of a simple buck converter, the chapter describes how loop analysis is insufficient in determining the conduction of power devices during switching events. The chapter then describes how nodal analysis can be used effectively in determining how power devices conduct and the transfer of current from one device to another. The chapter introduces the concept of events and how the matrix equations for the circuit will be constant until an event occurs. The chapter finally describes the logical flow of processes in the simulator as it performs loop analysis and nodal analysis one after the other.

Keywords Nodal analysis · Buck converter · DC–AC converter · Conduction state · Freewheeling · Circuit events · Simulation flow

8.1 Introduction

The previous chapter described loop analysis as the foundation of circuit analysis in the simulator. As stated, as the focus of the simulator is toward power electronics applications, this makes loop analysis readily applicable. Additionally, loop analysis serves as a tutorial for learning the workings of power electronic circuits by generating the closed paths that the current flows through. Chapter 6 presented the concept of the LoopMap and KCLBranchMap that contains the connectivity of the circuit and is used for performing circuit analysis. The simulator uses nodal analysis as a necessity under two conditions—to be able to determine the current through stiff branches and to be able to determine the conduction of nonlinear devices.

As stated in the previous chapter, loop analysis tends to be unstable while determining the current through stiff branches. This is due to the nature in which loop analysis deals with stiff branches—isolating them and approximating them. A far more stable and accurate manner of determining the currents through stiff branches would be to determine the voltage drops across them. This is where the simulator uses nodal analysis. Loop analysis results in the currents of the non-stiff branches

S. V. Iyer, *Simulating Nonlinear Circuits with Python Power Electronics*,
https://doi.org/10.1007/978-3-319-73984-7_8

being computed with a high degree of accuracy. These results of the loop analysis form the foundation of nodal analysis.

Nodal analysis as performed in the simulator is based on solving the circuit at a snapshot of a time instant instead of iteratively like loop analysis. Nodal analysis does not use the integration methods used in loop analysis to progressively compute branch currents with respect to a previous time instant. On the other hand, nodal analysis will use the branch currents and branch voltages generated from loop analysis to compute the node voltages and thereafter the currents in the stiff branches which are the only blanks to be filled in after loop analysis. Nodal analysis uses voltage sources, inductor currents, and capacitor voltages as driving inputs and is found to result in a fairly stable computation of node voltages while taking into consideration the effect of nonlinearity.

As this is the final chapter of the book, this chapter will deal with how power electronic circuits are dealt with in the simulator. The chapter will use a simple power converter to show how loop analysis is not completely sufficient to simulate nonlinear circuits. The example establishes the need for nodal analysis to determine the conduction of nonlinear devices after every switching event. In particular, nodal analysis is used to determine the freewheeling of inductor current and conversely the impact of the inductor current in preventing the conduction of some devices. The chapter will describe how the simulator uses an event to signal how a change in one or more branches results in the circuit being reprocessed with respect to LoopMap and KCLBranchMap. The chapter will also describe the sequence in which the simulator processes loop analysis and nodal analysis.

8.2 Concept of Nodal Analysis

The concept of nodal analysis used in this circuit simulator is different from the traditional approach of implementing nodal analysis. This is because nodal analysis plays a supportive role to the loop analysis. In the beginning, we will examine how nodal analysis is implemented subsequent to loop analysis at every time step of the simulation. Later in the chapter, we will describe how nodal analysis is used to determine the conduction of Switches and Diodes particularly with respect to freewheeling in a nonlinear circuit. Let us consider the circuit in Fig. 8.1. For simplicity, the above circuit does not have any nonlinear elements. The loop manipulations have isolated the stiff branch Voltmeter Vm to a single loop, i.e., Loop 3. Let us assume that loop analysis has been performed and the currents through all the branches have been calculated by the method of mapping branch currents and loop currents as described in the previous chapter. With this as our starting point, we will now describe how nodal analysis will be used to calculate the current in the branch having Voltmeter Vm.

The objective of nodal analysis will be to determine the voltages Vx1 and Vx2 at the nodes shown in the circuit. A detailed nodal analysis performed on the above circuit will involve the solving of a differential equation as the above circuit contains a capacitor and an inductor. However, as stated before, the nodal analysis is merely a

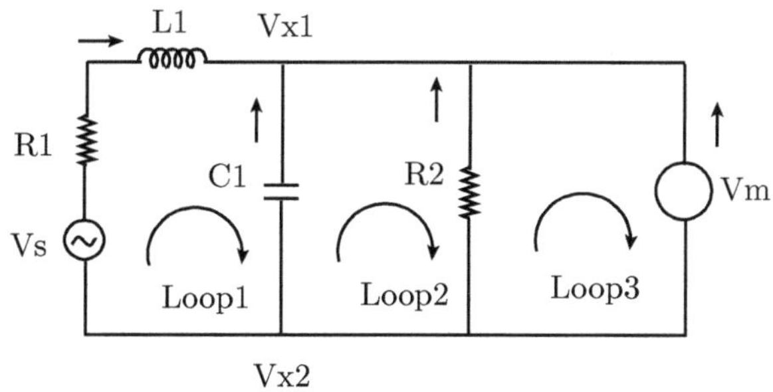

Fig. 8.1 Nodal analysis for a sample circuit

supplement to the loop analysis; and therefore, the nodal analysis will not be carried out in the same detail. The objective of the nodal analysis is to determine the current through the stiff branch which in this case is the branch with Voltmeter Vm. To perform a simplified nodal analysis, the following assumptions are made:

1. A branch with an inductor is considered to be a current source. This is due to the fact that the current through the inductor which has been calculated by the loop analysis will not change due to the nodal analysis. An exception to the rule would be when the branch with the inductor also has a large resistor making the branch a stiff branch. In that case, the inductor will be neglected and the branch will be treated as a resistive branch.
2. All capacitors are considered to be voltage sources. Again, this is due to the fact that the capacitor voltage is determined by loop analysis which calculates the current in a branch that has a capacitor and subsequently determines the capacitor voltage. Neither the current in the capacitor branch nor the voltage of the capacitor will change with the nodal analysis.

There are two reasons for choosing inductors as current source. First, as said before, the current through an inductor will not change instantaneously; and therefore, in nodal analysis, a branch with an inductor can be considered as a source that supplies or draws a current at or from a node. Second, if nodal analysis was to be performed without inductors as current sources, the only other option is to consider them as equivalent voltage drops. This would imply calculating Ldi/dt which can be fairly unstable as the di/dt particularly in nonlinear circuits with switching elements can be quite high and fluctuating.

In the circuit shown above, the branch {Vs, R1, L1} will be considered a current source provided R1 is not too large and makes the branch stiff. The other branches will be treated as normal resistive branches including the capacitor C1 which will be treated as a voltage source with a parasitic resistance. It should be noted that the currents in these branches have already been determined with loop analysis. However, they are still considered in the nodal analysis as resistive branches instead of mere current sources. This is due to the fact that if all the branches except the stiff branch are considered as current sources, the calculation will involve determining the flow of the net current of all these branches into the stiff branch. For the above simple circuit, this might be possible, but for larger circuits, this is error prone as the matrices with only stiff branches will be ill-formed and the resultant node voltages will be unstable. On the other hand, even though the non-stiff branches are considered as resistive branches, the currents through them are not updated after the nodal analysis but the values obtained from loop analysis are retained. This is to minimize the computational burden by calculating the currents only through the stiff branches.

The nodal analysis involves solving static equations. The matrix equation for nodal analysis can be written as:

$$\mathbf{Yx} = \mathbf{I} \tag{8.1}$$

The vector **x** is a collection of the voltages at the nodes. At this point, it needs to be specifically stated that the nodes considered exclude the short nodes that were described in the previous chapter. For example, in the circuit of Fig. 8.1 taken as an example, there are four nodes. However, the two nodes at the top are connected together by a wire making their voltages to be the same and the same goes for the two nodes at the bottom. The node used to represent such a group of nodes connected by wires is chosen arbitrarily among them and is called a KCL node as this is the node whose voltage is calculated by nodal analysis. A detailed explanation of this concept can be found in Chap. 6 where short nodes and KCLBranchMap are presented. As done for the loop analysis, we will start by writing the equations for nodal analysis and generalize it for larger circuits. Let us consider a circuit with three nodes as the one in Fig. 8.2. Applying KCL at the three nodes whose voltages are marked as Vx1, Vx2, and Vx3 produces the following equations.

$$-I_{L1} + \frac{V_{x1} - V_{c1} - V_{x3}}{R_{c1}} + \frac{V_{x1} - V_{x2}}{R_3} = 0 \tag{8.2}$$

$$-\frac{V_{x1} - V_{x2}}{R_3} + \frac{V_{x2} - V_{x3}}{R_2} + \frac{V_{x2} - V_{x3}}{R_{Vm}} = 0 \tag{8.3}$$

$$I_{L1} - \frac{V_{x1} - V_{c1} - V_{x3}}{R_{c1}} - \frac{V_{x2} - V_{x3}}{R_2} - \frac{V_{x2} - V_{x3}}{R_{Vm}} = 0 \tag{8.4}$$

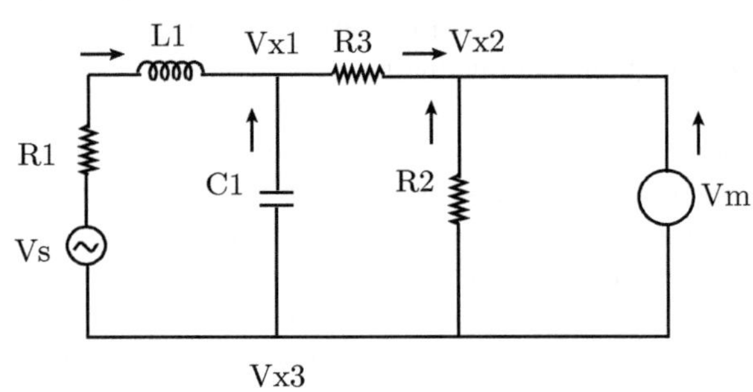

Fig. 8.2 An expanded circuit

The current entering a node is considered negative while the current leaving a node is considered positive. The converse rule could also be applied with the same results. The equations can be written in matrix form as follows

$$\begin{bmatrix} \frac{1}{R_{c1}} + \frac{1}{R_3} & -\frac{1}{R_3} & -\frac{1}{R_{c1}} \\ -\frac{1}{R_3} & \frac{1}{R_2} + \frac{1}{R_3} + \frac{1}{R_{Vm}} & -\left(\frac{1}{R_2} + \frac{1}{R_{Vm}}\right) \\ -\frac{1}{R_{c1}} & -\left(\frac{1}{R_2} + \frac{1}{R_{Vm}}\right) & \frac{1}{R_{c1}} + \frac{1}{R_2} + \frac{1}{R_{Vm}} \end{bmatrix} \begin{bmatrix} V_{x1} \\ V_{x2} \\ V_{x3} \end{bmatrix} = \begin{bmatrix} I_{L1} + \frac{V_{c1}}{R_{c1}} \\ 0 \\ -I_{L1} - \frac{V_{c1}}{R_{c1}} \end{bmatrix} \tag{8.5}$$

A few observations about the matrix equations above. The matrix **Y** is symmetrical. This is because the off-diagonal terms are the admittances of the branches connecting two nodes. Therefore, the off-diagonal terms (1, 2) and (2, 1) are the admittance 1/R3 of the branch connecting nodes with voltages Vx1 and Vx2, respectively. Secondly, the diagonal terms are always positive while the off-diagonal terms are negative. This is because of the manner in which each branch current can be expressed when writing the nodal equations. For example, consider Eq. 8.3 at node with voltage Vx2. All the currents entering node Vx2 are expressed as voltage Vx2 with respect to the other node voltages Vx1 and Vx3. The first branch current entering Vx2 is expressed as (Vx1 − Vx2)/R3 but is taken with a negative sign as this is the current entering the node Vx2. Therefore, the term associated with node Vx2 due to the current in the branch R3 is still +1/R3. The second term is the current through branch R2 which is exiting the node Vx2 and is (Vx2 − Vx3)/R2, and since this current is exiting, it is positively resulting in a term +1/R2 associated with node Vx2. This implies that whatever may be the direction of current assumed while performing nodal analysis, the diagonal terms will always be positive. Conversely, the off-diagonal terms will always be negative which can be deduced from examining the nodal equations.

As described in Chap. 7 on loop analysis, the above matrix equation can be transformed to an upper triangular form using elementary row operations. The transformed matrices will appear in the following form.

$$\begin{bmatrix} y_{11} & y_{12} & \cdots & y_{1N} \\ 0 & y_{22} & \cdots & y_{2N} \\ \vdots & \vdots & \ddots & \vdots \\ 0 & 0 & \ldots & y_{NN} \end{bmatrix} \begin{bmatrix} V_1 \\ V_2 \\ \vdots \\ V_N \end{bmatrix} = \begin{bmatrix} I'_1 \\ I'_2 \\ \vdots \\ I'_N \end{bmatrix} \tag{8.6}$$

The above matrix equations can be solved backwards in the order Vn, Vn-1,....., V1. The original matrix **Y** has been generated by applying KCL at all the nodes in the circuit. A reference or a ground node has not been assigned in the circuit. Typically, while applying nodal analysis, one of the nodes in the circuit is the ground node or a node is assumed to play the role of the ground or reference node and have a voltage of 0. While analyzing a circuit, a human can assign a node as a reference node based on the location of the node. However, to program this choice of reference node is extremely difficult with only the connectivity information of the circuit. The other option would be to force the user to assign a node as ground. The problem with this method is that a node in a circuit which is not anywhere near 0 V might end up being called a ground node just for the purpose of analysis. Furthermore, in the case of systems with transformers or isolated electrically by other means, assigning ground nodes in each subcircuit could be troublesome as this might also disrupt the concept of an isolated system. For this reason, the circuit simulator does not assume a ground node or expect the user to assign one. By applying KCL at all nodes in the circuit, the matrix **Y** in the KCL equation, becomes singular. Therefore, the row operations performed on the nodal matrix equation will result in the last diagonal term ynn to be zero and, therefore, the last row to be zero. For this reason, the voltage of the last node in the circuit is zero. Since, what interests us is the voltages across the branches, one of the nodes being at zero voltage for the sake of analysis does not pose a problem.

Once all the voltages Vn, Vn-1, …, V1 are determined, the voltages at the KCL nodes have been determined. The next step is to use these KCL node voltages to determine the voltages at all the short nodes. The final step is to be able to calculate the currents in the branches. As stated before, the nodal analysis is used to determine the currents only through the stiff branches. With this description of the basic concept of nodal analysis, we shall now proceed to describe how nodal analysis is used to determine the state of a nonlinear circuit after the change in the state of one or more components.

8.3 Limitation of Loop Analysis in Nonlinear Circuits

The previous chapter described a limitation of loop analysis to be able to calculate the currents through stiff branches accurately. However, nodal analysis becomes critical in determining the conduction of nonlinear elements in power electronic circuits. As an example, let us consider a buck converter in Fig. 8.3. The circuit has three possible conduction states as shown below the circuit. When the Diode or the Switch is in off state, they are high resistances such that at the maximum specified voltage, they would draw a negligible current. Before jumping into how the circuit simulator deals with this, a basic description of how the converter works [5]:

1. Switch S1 is turned on while the Diode is on in which case the Diode turns off as it is reverse biased by the source Vin. If Switch S1 is turned on when Diode is off, the Diode is reverse biased by the source Vin and remains in the off state. The input dc source charges the inductor and the output capacitor resulting in a rise in both the inductor current and the capacitor voltage.
2. Switch S1 turns off. The current through the inductor now freewheels through the Diode D1 and the inductor current gradually decreases. The capacitor supplies the load and the capacitor voltage also decreases.
3. The current decreases to zero, when the Diode also turns off. This state may not occur as the Switch S1 may turn on again before the current decreases to zero.

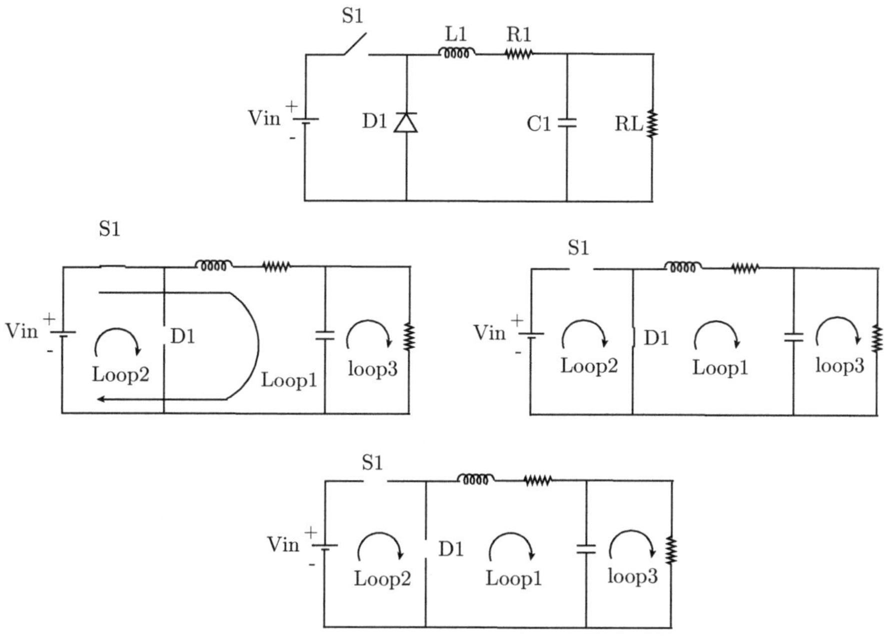

Fig. 8.3 Buck converter as an example of a nonlinear circuit

Though it may seem obvious to anyone well-versed with power electronics, this is the beginning of a challenge with respect to the circuit simulator—how will current freewheel in the presence of multiple nonlinear elements in a circuit? Let us examine the transition—S1(on)-D1(off) to S1(off)-D1(on) in Fig. 8.3. When the Switch S1 gets the gate turn off signal, it transitions to a high resistance in the next simulation instant. If the loop analysis were to continue as usual, the current in loop 1 would become negligible. The current in loop 2 was negligible to begin with. The loop manipulations would not help to able to determine if the Diode D1 was to conduct because if you examine the second figure, loop 2 would have a current such that D1 would be reverse biased by the voltage source Vin. Therefore, according to loop manipulations, the current in the inductor L1 would become negligible and would remain so until the Switch S1 conducts again. To understand the concept of freewheeling, we need to answer the question why does freewheeling take place? In the circuit of Fig. 8.3, freewheeling would take place as the current in the inductor cannot change instantaneously. Any change in the inductor current would result as per Lenz's law in an induced emf which seeks to oppose the cause which produces it. To depict this, take a look at Fig. 8.4. The figure on the left shows how the induced emf of the inductor is such that it opposes the current which is increasing when the Switch is closed. On the other hand, as the Switch begins to open and the resistance starts to increase, the drop in current causes the induced emf to change such that it opposes this change. Therefore, the direction reverses in order to oppose this change (decrease) in current. When the induced emf exceeds the capacitor voltage due to

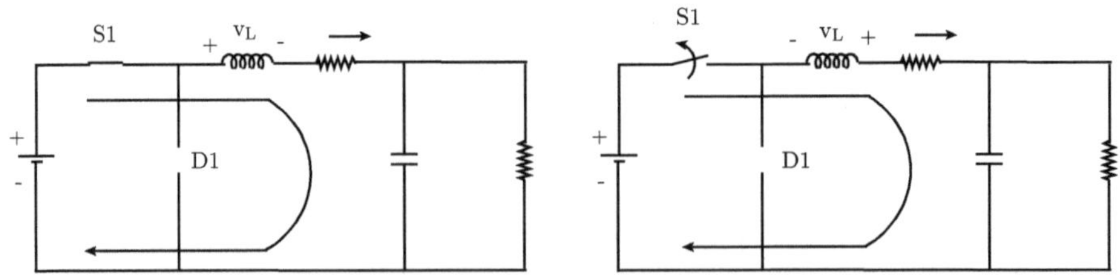

Fig. 8.4 Freewheeling of an inductor

dropping current, the Diode D1 becomes forward biased and begins to conduct. The inductor current now freewheels through the Diode D1 and continues to supply the capacitor C1.

Several attempts were made to modify loop analysis to be able to achieve freewheeling. However, the fundamental problem in all these cases was that it was the inductor current that was attempting to find another path to continue flowing. Loop analysis is essentially a voltage summation law which states that the sum total of source voltages and resistive drops must equal zero in a loop. This in turn translates to being able to model the inductor as a voltage source. Though it is possible to replace the inductor by a voltage source of magnitude Ldi/dt in steady state, the question arises—how will this help in transients? For example, when the Switch begins to open, in actual practice, the almost imperceptible drop in current is sufficient to generate an induced emf that causes freewheeling. However, the time frame for this transition is extremely narrow. A MOSFET or an IGBT may turn off in less than a microsecond. This is the time for the current through it to decrease to zero when the gate signal to turn off has been applied. The time interval for freewheeling to begin would be an extremely small fraction of that—in all probabilities in nanoseconds or picoseconds. To simulate this digitally would be extremely difficult as it may need a variable time step that changes when a transition occurs. Though it may be feasible for small circuits, for larger circuits, this is a huge computational burden.

The solution to this problem is to consider the inductor current as an input source and to examine through which branch this could potentially flow. Let us consider again the circuit in Fig. 8.5 to demonstrate this concept. The current I_{L1} can be supplied by either the Switch current I_{S1} or the Diode current I_{D1}. Therefore, assuming for convenience that they have equal off resistances, these two currents I_{S1} and I_{D1} would be equal and half of I_{L1}. The next question to ask would be—which of these branches might be able to change their state due to this new current calculated due to freewheeling? The Switch S1 has been turned off by a gate signal and therefore will stay turned off. As for the Diode, if a current I_{D1} flows in the direction shown in the figure, it would become forward biased and turn on. Therefore as a concept, this is a solution to determining the freewheeling of nonlinear devices. Since this concept requires the calculation of branch currents with respect to another branch current, the only way to perform this systematically for a large circuit is through nodal analysis as described in the previous section. This entire process of determining freewheeling will be described in the following sections.

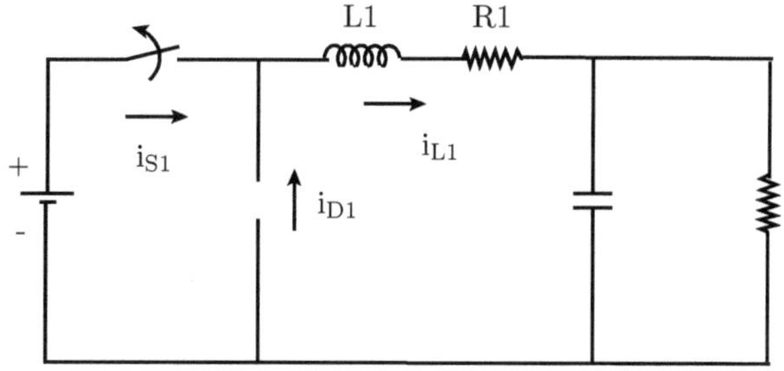

Fig. 8.5 Inductor as a current source to achieve freewheeling

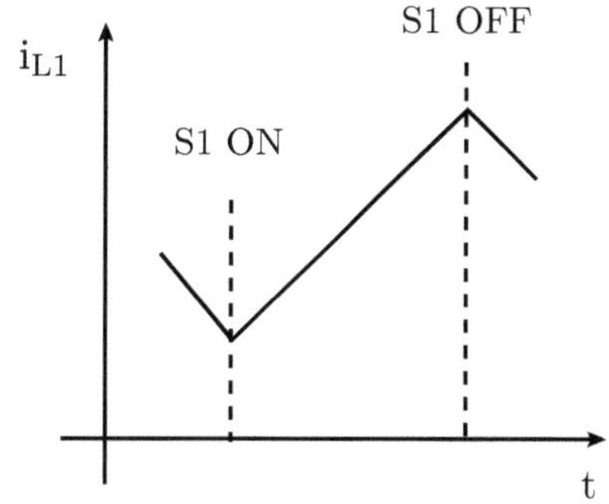

Fig. 8.6 Current with Switch state

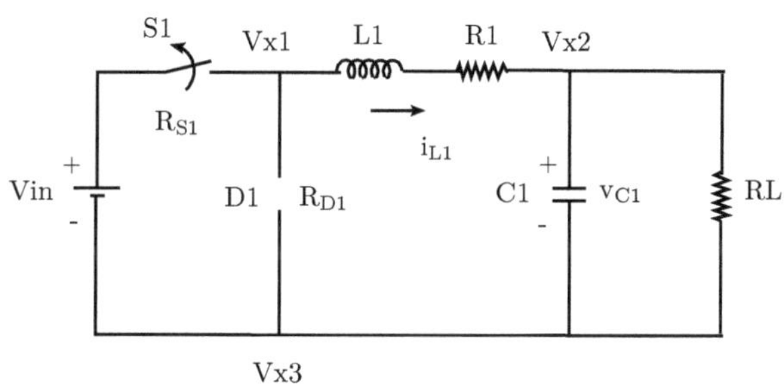

Fig. 8.7 Circuit when Switch turns off

8.4 Applying Nodal Analysis in Nonlinear Circuits

The method of applying nodal analysis is almost identical to that described in the beginning of this chapter. However, to illustrate how nodal analysis helps to determine the state of conduction of devices, the example of the buck converter will be considered. To begin with let us consider the first state of the circuit, i.e., when the Switch S1 is closed and the Diode D1 is reverse biased and off. Figure 8.6 shows how the current I_{L1} increases when the Switch S1 is turned on until the moment it is turned off. The circuit at the moment when Switch S1 is turned off is as shown in Fig. 8.7. The nodal matrix equation for the above circuit will be:

$$\begin{bmatrix} \frac{1}{R_{S1}} + \frac{1}{R_{D1}} & 0 & -\frac{1}{R_{S1}} - \frac{1}{R_{D1}} \\ 0 & \frac{1}{R_{c1}} + \frac{1}{R_L} & -\frac{1}{R_{c1}} - \frac{1}{R_L} \\ -\frac{1}{R_{S1}} - \frac{1}{R_{D1}} & -\frac{1}{R_{c1}} - \frac{1}{R_L} & \frac{1}{R_{S1}} + \frac{1}{R_{D1}} + \frac{1}{R_{c1}} + \frac{1}{R_L} \end{bmatrix} \begin{bmatrix} V_{x1} \\ V_{x2} \\ V_{x3} \end{bmatrix} = \begin{bmatrix} -I_{L1} + \frac{V_{in}}{R_{S1}} \\ I_{L1} + \frac{V_{c1}}{R_{c1}} \\ -\frac{V_{in}}{R_{S1}} - \frac{V_{c1}}{R_{c1}} \end{bmatrix} \tag{8.7}$$

In order to solve the above equation, let us assume some values for the circuit parameters as follows:

```
Vin  = 100V,
L1  = 1 millihenry
R1  = 0.1 ohm
RL  = 100 ohm
C1  = 1000 microfarad
```

At the moment when S1 is turned off, let us assume that the inductor current I_{L1} and the capacitor voltage V_{c1} are:

```
IL1  = 10A
Vc1  = 50V
```

Moreover, let us assume that the off resistance of the Diode and the Switch are 1 megaohm. So,

```
Rs1  = 1 megaohm
Rd1  = 1 megaohm
```

The simulator will use the above values to solve the matrix equation and arrive at the nodal voltages V_{x1}, V_{x2}, and V_{x3}. The purpose of this nodal analysis is to determine the currents through any branches with nonlinear elements which in this case will be the branch {Vin, S1} and D1. Therefore, the simulator will calculate these currents as:

$$I_{S1} = \frac{V_{x1} - V_{in} - V_{x3}}{R_{S1}} \tag{8.8}$$

$$I_{D1} = \frac{V_{x1} - V_{x3}}{R_{D1}} \tag{8.9}$$

Let us arrive at approximate values of these currents without having to calculate the entire nodal matrix equation. The current I_{L1} will divide into the branches {Vin, S1} and D1. Let us assume that the nodal voltage V_{x3} is zero, i.e., we assume this to be our reference node. In that case, the KCL equation at node V_{x1} will be:

$$\frac{V_{x1} - V_{in}}{R_{S1}} + \frac{V_{x1}}{R_{D1}} = -I_{L1} \tag{8.10}$$

Substituting the values chosen above:

$$\frac{V_{x1} - 100}{1e + 6} + \frac{V_{x1}}{1e + 6} = -10 \tag{8.11}$$

which produces $V_{x1} = -4999950$ V. This in turn produces the currents $I_{S1} = 5.00005$ A and $I_{D1} = 4.99995$ A. The next question to be asked is—if these currents were to flow through these branches, would there be any changes in the state of the devices in them? Since Switch S1 is controlled by the gate signal, once it turns off, it remains off. On the other hand, current $I_{D1} = 4.99995$ A flowing in the direction shown, would result in the Diode turning on. In this case, the Diode does not turn on in the conventional manner as that requires that the Diode be forward biased by a voltage applied across it when it exceeds its threshold voltage of approximately 0.7–1 V. For this reason, to accommodate this special condition where a Diode can be turned on by a current instead of a voltage, the Diode class is provided with a special function to determine if it can freewheel. In this function, it is checked whether the voltage drop produced by the product of the current flowing through the Diode obtained from nodal analysis and the resistance of the Diode at that simulation instant is greater than the threshold voltage. Since 4.9995*1e+6 is quite clearly larger than 1 V, the Diode is determined to turn on due to the current through the inductor.

Before continuing with how the simulator will process the above information, let us also consider the other case—when S1 turns on with Diode D1 freewheeling. The waveform and the circuit are shown in Fig. 8.8. We will skip writing the nodal matrix equations and instead examine how the currents through the branches with nonlinear elements will be during the transition. Most of the values of the circuit parameters will be the same as the previous case. Let us assume that the on resistance of the Diode D1 and the Switch S1 are 0.01 ohm. Let us also assume that at the instant of of Switch S1 turning off, inductor current $I_{L1} = 2$ A. The KCL equation at node V_{x1} can be written as:

$$\frac{V_{x1} - 100}{0.01} + \frac{V_{x1}}{0.01} = -2 \tag{8.12}$$

which produces $V_{x1} = 49.99$ V. Currents are calculated as $I_{S1} = 5001$ A and $I_{D1} = -4999$ A. As before, the question asked is—would these currents produce a change in the state of devices within these branches? The Switch S1 has received a turn-on signal at the gate and from the direction and magnitude of the current I_{S1}, S1 will begin to conduct. On the other hand, the current $I_{D1} = -4999$ A will cause the Diode

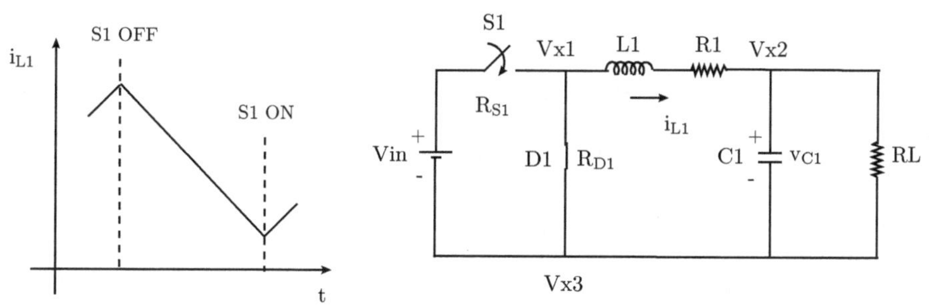

Fig. 8.8 Circuit when Switch turns on

to turn off as a Diode does not conduct in the reverse direction. Therefore, as in the previous case, the nodal analysis will help to determine which nonlinear elements will change their state.

The two states of the buck converter show that nodal analysis can determine when nonlinear elements can turn on and turn off based on the current through the inductor. The strategy is also applicable in larger and more complex circuits. Performing nodal analysis helps to determine the effect of an inductor current even when the nonlinear element is not in a branch incident at the same node as the inductor. The next section will describe how the simulator incorporates the above nodal analysis with the loop analysis.

8.5 Continuing with Loop Analysis

The previous section described the concept of nodal analysis to determine the conduction of nonlinear devices. However, the core engine of the simulator remains the loop analysis. Therefore, the question—how do we translate the above-determined conduction state of the circuit into the next iteration of loop analysis? There are two parts of the problem—the first being generating an accurate branch current map of the circuit after the new conduction state of the circuit has been determined. And second, generate a new LoopMap for this circuit and calculate the values of the loop currents.

As seen in the previous section, when the nodal analysis is performed to determine the currents through nonlinear devices for the currents through inductors, the currents obtained are not the actual currents. For example, consider the first case in the above example when Switch S1 is turned off. The nodal analysis produces the currents $I_{S1} =$ 5.00005 Amps and $I_{D1} = 4.99995$ Amps. These currents merely help to determine if any of the nonlinear devices will change their state. In this case, it was found that the Diode would change its state due to $I_{D1} = 4.99995$ Amps. However, this would not be the actual current for the simple reason that with the Switch S1 off, I_{S1} cannot be equal to 5.00005 Amps. With S1 turned off and remaining off, the current I_{S1} must be negligible as the turn off resistance of the Switch S1 will be in the order of megaohms. Moreover, almost the entire inductor current $I_{L1} = 10$ Amps would flow through the Diode except for this negligible current through the Switch. The second case when the Switch S1 turns on while Diode D1 is conducting is even more obvious. Nodal analysis produces currents $I_{S1} = 5001$ Amps and $I_{D1} = -4999$ Amps which are quite obviously not the currents that would flow when the inductor current is only 2 Amps. Thus, the nodal analysis described in the previous section only serves to determine the state of conduction of the circuit.

To elaborate on this further, a power electronics circuit with several nonlinear elements will receive a number of control signals simultaneously or may have elements changing their state for a number of reasons. With every change or attempt to change the state of conduction of the circuit, it needs to be decided whether the state of the circuit will allow certain changes to occur. To simplify the above fairly complex argument, let us consider a single-phase voltage source inverter in Fig. 8.9.

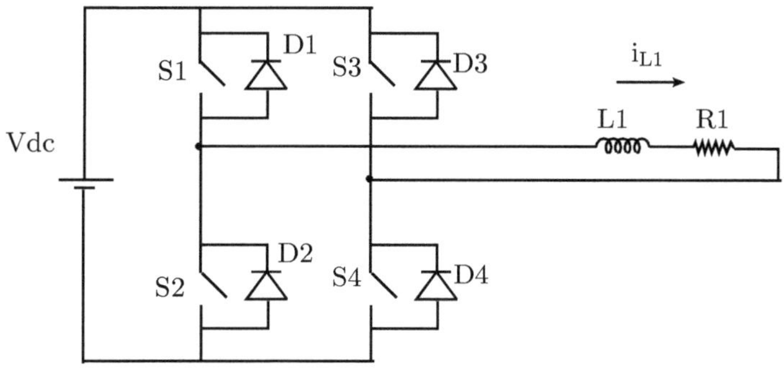

Fig. 8.9 Single-phase inverter

Fig. 8.10 Effect of inductor current on Switch state

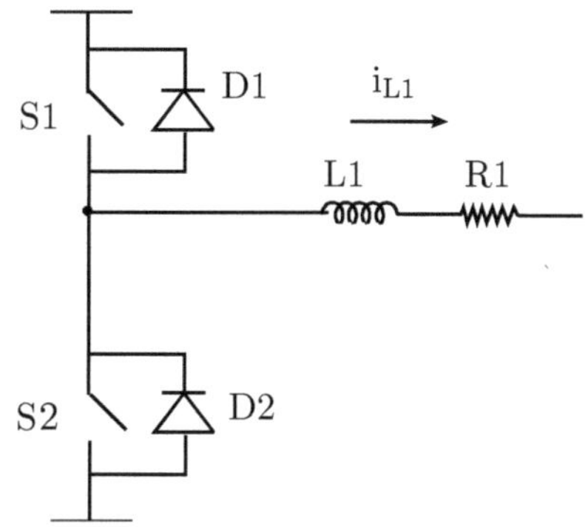

Let us assume a simple pulse width modulation scheme is used to generate the gate signals for the four Switches S1 to S4. Let us consider the case when Switches S1 and S4 are turned on and all other devices are turned off. The current I_{L1} indicated in the circuit will rise. Let us assume that this current I_{L1} is positive. Soon after, S1 and S4 receive gate signals to turn off and S2 and S3 receive gate signals to turn on. For a Switch to turn on it needs to satisfy two conditions—it must be forward biased and it must conduct current in the positive direction only. Examine the diagram in Fig. 8.10 showing Switches S1 and S2 with their anti-parallel Diodes D1 and D2. The Switch S2 is forward biased as even if the Switch S1 is turned off and is high resistance, the voltage drop across the two Switches will be equal to Vdc/2 assuming they have equal turn off resistances. The next condition fails—from the direction of inductor current I_{L1}, if the Switch S2 conducts, it will have to conduct a current in the reverse direction which it cannot. As before, the current through an inductor cannot change instantaneously. Since Switch S1 will turn off since it has received a turn off signal at its gate, and S2 cannot conduct, the Diode D2 will conduct and the current I_{L1} will flow through it. This is precisely the reason why anti-parallel Diodes are used in a voltage source inverter. This is an example where a Switch has received a turn-on signal at its gate but finds that it cannot conduct because the inductor at its output will not let it do so. As with the example of the buck converter, nodal analysis can be used to determine which device will conduct and which device will stay turned off.

Once the conduction state of the circuit is determined, it is needed to update the currents in all branches of the circuit. At this point, it should be stressed that currents in all branches need to be recalculated as the nonlinear elements that have been determined to have turned on or off could have changed the currents in any or all branches in the circuit. The only exceptions are the currents in branches that have inductances. This is because the current through an inductance cannot change instantaneously, and therefore, these branches are treated as current sources in the upcoming nodal analysis. Let us again consider the buck converter above as an example. The waveforms and the circuit are shown together in Fig. 8.11 [5]. In order to determine the branch currents at the two events nodal analysis can be performed using the matrix equation:

$$\begin{bmatrix} \frac{1}{R_{S1}} + \frac{1}{R_{D1}} & 0 & -\frac{1}{R_{S1}} - \frac{1}{R_{D1}} \\ 0 & \frac{1}{R_{c1}} + \frac{1}{R_L} & -\frac{1}{R_{c1}} - \frac{1}{R_L} \\ -\frac{1}{R_{S1}} - \frac{1}{R_{D1}} & -\frac{1}{R_{c1}} - \frac{1}{R_L} & \frac{1}{R_{S1}} + \frac{1}{R_{D1}} + \frac{1}{R_{c1}} + \frac{1}{R_L} \end{bmatrix} \begin{bmatrix} V_{x1} \\ V_{x2} \\ V_{x3} \end{bmatrix} = \begin{bmatrix} -I_{L1} + \frac{V_{in}}{R_{S1}} \\ -I_{L1} + \frac{V_{c1}}{R_{c1}} \\ -\frac{V_{in}}{R_{S1}} - \frac{V_{c1}}{R_{c1}} \end{bmatrix} \tag{8.13}$$

When Switch S1 is turned off and Diode D1 starts to freewheel, R_{D1} will be equal to the on resistance of the Diode while R_{S1} will be equal to the off resistance of the Switch. I_{L1} and V_{c1} will be the inductor current and the capacitor voltage at the instant of switching. Similarly, when Switch S1 is turned on and Diode D1 turns off, R_{D1} will be equal to the off resistance of the Diode while R_{S1} will be equal to the on resistance of the switch. Calculation of the node voltages V_{x1}, V_{x2}, V_{x3} can be used in the calculation of all the branch currents of the circuit except for the branch with the inductor.

From the waveforms, it can be seen that there is a discontinuity in both the Switch current I_{S1} and the Diode current I_{D1} when the Switch S1 turns on or off. In this

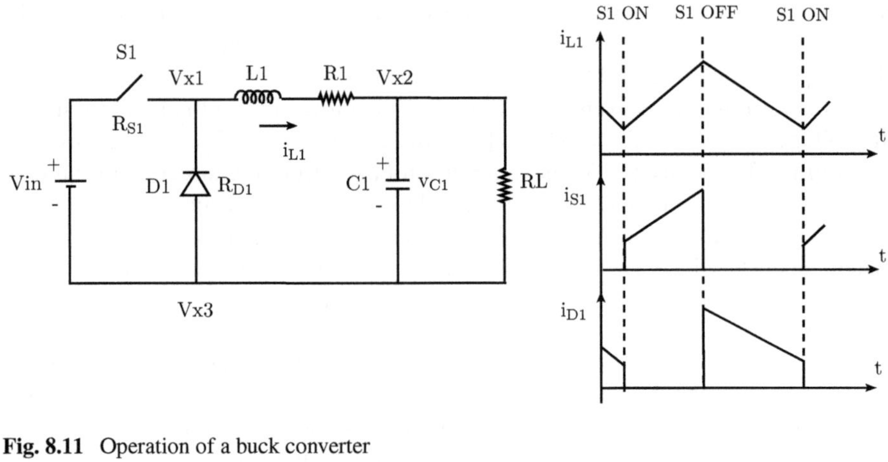

Fig. 8.11 Operation of a buck converter

circuit, it is a coincidence that only the currents in branches with nonlinear elements experience a discontinuity. However, in more complex circuits, it is always possible that a discontinuity appears in any branch except of course for the branches with inductances as these are current sources. Typically, a discontinuity will appear in any branch that is incident at a node which also has another branch incident at it with a nonlinear element undergoing a change in state. This is for the simple reason that as per Kirchhoff's Current Law, the sum of currents at every node has to be zero, and therefore, a sudden change in the current through one branch incident at a particular node will affect the current through all branches incident at the same node. Therefore, it is essential to use nodal analysis in this case to update the currents in all branches of the circuit except for the branches with inductances.

Nodal analysis can thus be used to determine the manner in which current is transferred between branches when nonlinear elements change their state. However, to continue with the simulation, the simulator needs to return to loop analysis. A few changes have occurred due to the change in states of nonlinear elements. The parameters of branches have changed—quite specifically in the field of power electronics, some branches that were very large resistances in the order of megaohms, and therefore, stiff may now be negligible resistances in the order of milliohms. The reverse is also possible—branches with negligible resistance may now be stiff branches with large resistances. As a result, LoopMap needs to be updated with the latest information of which branches are stiff and which are non-stiff. And, as described in the previous chapter, it may be necessary to perform loop manipulations in order to restrict some stiff branches to the minimum number of loops. This is because non-stiff loops formed using some branches that were previously non-stiff may now become stiff if these branches had nonlinear elements that turned off. Using loop manipulations, the non-stiff loops can be extracted and be fed to the differential equation solver. However, the circuit is not at "rest." Before, in the previous chapter, when loop manipulations were performed to restrict the Voltmeter to the minimum number of loops, the simulation had not started and therefore all branch currents and loop currents were zero, i.e., the circuit was at "rest." Now the branch currents are not zero and neither are the loop currents. The loop currents that were obtained as the solution of the differential equation in the previous iteration were associated with a completely different LoopMap. Therefore, to be able to solve the new differential equations that are obtained from the new LoopMap, the loop currents will need to be calculated for this new LoopMap. The only way to do so is using the branch currents. This method has been described in the previous chapter where it was shown how the LoopMap established a relationship between the loop currents and the branch currents. The method of calculating loop currents from the branch currents was described in detail in Chap. 7.

Now that the loop currents have been calculated and the new differential equations are obtained from the LoopMap, the method of solving them as shown in the previous chapter can be used. The next question that arises is—when do these circuit updates need to be done? To answer this, we introduce the concept of event-driven circuit updates.

8.6 Event-Driven Circuit Updates

Before describing the concept of events in a power electronics circuit, let us examine how a passive circuit is solved. Let us consider the passive circuit in Fig. 8.12. The circuit contains at least one stiff element, i.e., the Voltmeter Vm. Depending on the value of resistances R1 and R2, those branches could also be stiff. After reading the circuit from the spreadsheet, the nodes, branches, and loops are determined. Subsequently, the BranchMap, LoopMap, and KCLBranchMap are determined in order to perform loop analysis and nodal analysis. The LoopMap needs to be processed with loop manipulations to restrict the stiff branches to the minimum number of stiff loops and therefore extract the non-stiff loops. In order to ensure that this is done, before starting the simulation time increments, even a passive circuit such as the one above is considered to generate an event. An event in a circuit occurs when a parameter of any element in any branch of the circuit changes. By parameter, it is implied the resistance or inductance of an element because a change in the voltage of a source will be considered as an input to the system and not a parameter. At the beginning, every branch in the circuit carries an event flag to ensure that the circuit is processed. Once, LoopMap is processed, all the matrices **E**, **A**, and **B** are updated in the differential equation for loop analysis.

$$\mathbf{E}\frac{d\mathbf{x}}{dt} = \mathbf{Ax} + \mathbf{Bu} \tag{8.14}$$

Furthermore, the matrix **Y** is updated in the nodal matrix equation.

$$\mathbf{YV} = \mathbf{I} \tag{8.15}$$

The circuit can be simulated by solving the above matrix equations at a constant time step and at each time step updating the voltages and currents that are inputs to these equations. No further events are generated for this passive circuit in Fig. 8.12, and therefore, LoopMap need not be subjected to loop manipulations.

Now let us consider a buck converter. This circuit can have a maximum of three states as shown before—{S1 on, D1 off}, {S1 off, D1 on}, and {S1 off, D1 off}.

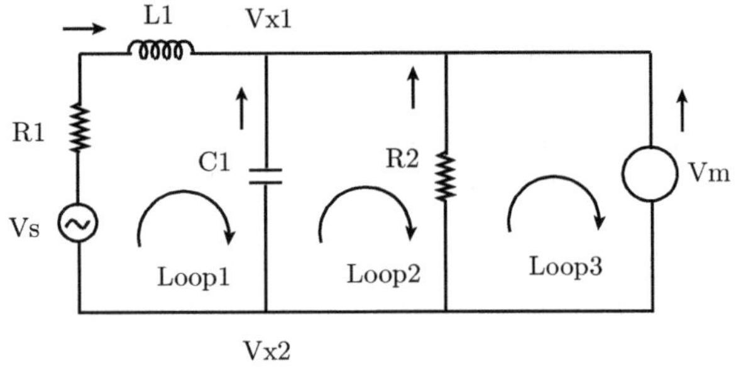

Fig. 8.12 A sample passive circuit

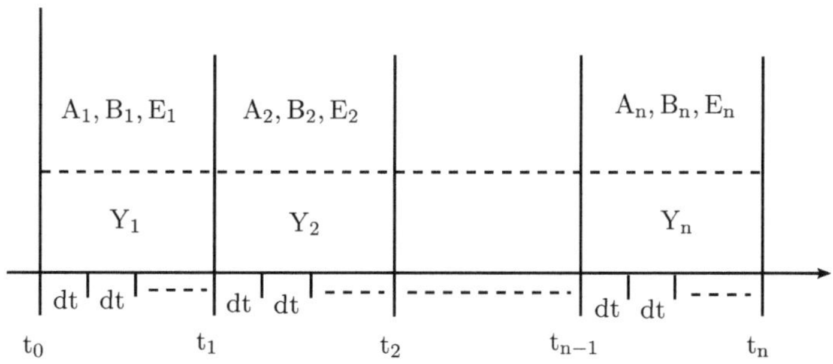

Fig. 8.13 Change in the state of the circuit

It may so happen that the third state never occurs if the converter is controlled in continuous conduction mode. Figure 8.13 shows any random nonlinear circuit being simulated a time step of dt for a time limit until tn. Depending on the nature of the circuit and the control applied to the elements, let us assume that the circuit changes its state at time instants (t1, t2, . . ., tn-1, tn). These time instants need not be regularly spaced which means that it could be possible that t2=t1+3*dt while t6=t5+10*dt. Between the time instants, there is no change in the parameters of the circuit, and therefore, the matrices **A**, **B**, and **E** in the loop equation and the matrix **Y** in the nodal equation remain constant. For a buck converter, there could be a maximum of three possible states—(**A1**, **B1**, **E1**, **Y1**), (**A2**, **B2**, **E2**, **Y2**), and (**A3**, **B3**, **E3**, **Y3**). For a buck converter, the states appear in a sequence due to the Diode taking up the Switch current when the Switch turns off and vice versa. In general, in a nonlinear circuit, any of these states could be repeated any number of times and could appear in any sequence because if they are controlled by user-defined control, they could be in a random sequence.

The first question—when and how are the events generated? An event is generated when any element in any branch experiences a change in resistance or inductance. This change could be small, for example, a load resistor gradually increasing in value as it heats up, or could be drastic such as a Diode turn off in which case its resistance will change from milliohms to megaohms. In any case, the element that experiences the change either due to its characteristic (Diode or heated resistor) or due to control (Switch) will trigger an event for the branch in which it is present. When an event is generated in one or more branches in the circuit, this means the circuit parameters have changed. This in turn means that the matrices (**A**, **B**, **E**, **Y**) in loop and nodal equations will change. The second question—what happens when an event has not been generated? This means that the circuit parameters have not changed and if the simulation time instant falls in any given time window (tm-1, tm), the matrices for the loop and nodal equation will be (**Am**, **Bm**, **Em**, **Ym**). Only the input vectors **u** and **I** containing the source voltages and the inductor currents will change.

The advantage of defining events is that for complex circuits, calculating the matrices (**A**, **B**, **E**, **Y**) continuously could be significant computational burden. To speed up the simulation, the calculation of loop and nodal analysis matrices can be limited to those times when there is an event. The next section will now describe how all these functions are scheduled in the simulator.

8.7 Process Flow in the Simulator

Combining loop analysis and nodal analysis in a circuit simulation might be a bit confusing. This would be further compounded by the fact that nodal analysis is used to determine the conduction of nonlinear elements and to determine the currents in stiff branches. This section will therefore list the processes in the circuit simulator to describe how the computational flow occurs. This would help to understand the importance of different functions and how they play a role in the simulator.

As stated before, when the circuit is read from the spreadsheet, the determination of branches, nodes and loops is performed and is followed by generating the BranchMap, LoopMap, and KCLBranchMap for the circuits. At the beginning of the simulation, all branch and loop currents are initialized to zero and usually all nonlinear elements are turned off. The initial state of the system can be obtained in a number of different ways, and it is always possible to perform a preliminary calculation to determine initial system conditions if a user specifies the initial conditions on some parameters like for instance inductor currents and capacitor voltages. However, this circuit simulator starts with the circuit "at rest" which in this case, means all currents to be zero. As stated in the previous section, at the beginning of the simulation, events are raised in each branch of the circuit. Due to this, the event-driven functions are processed. These are described in sequential order as follows:

1. The event has been generated because of changes in parameters of one or more branches in the circuit. When an event occurs in a circuit, it could be due to the switching of a nonlinear power electronic device. As explained before in the chapter, subsequent to an event, the conduction of devices will change and to examine which devices will start conducting or which devices will stop conducting, nodal analysis is performed. In the case of a passive circuit with one variable resistor, there will not be any other element changing its parameter except for this variable resistor. On the other hand, for a nonlinear power electronic converter, several elements may change their state. If no event has been generated, nodal analysis is not performed as there is not expected to be any change in the state of conduction of the circuit.
2. Once the state of conduction of the circuit has been determined, the parameters of all the branches in terms of its resistance and inductance can be updated. A change in the parameters of the circuit will result in a change in the flow of currents in the circuit. To calculate the currents in the branches, nodal analysis is used with the inputs being the voltage sources, capacitor voltages, and the inductor currents. All branch currents are calculated except for the currents in branches with inductances as the assumption made has been that the current through an

inductor cannot change instantaneously. If no event has been generated, this computation is also not performed, as no change the parameters of the circuit is expected which in turn will not cause a change in the branch currents from the previous iteration.

3. Since the parameters of the branches in the circuit have changed, it is possible that branches that were non-stiff in the previous simulation time step are now stiff while some branches that were stiff in the previous iteration are now non-stiff. LoopMap is updated with this latest information of the stiff and non-stiff branches in the circuit and loop manipulations are performed to restrict stiff branches to the minimum possible number of loops and extract the maximum possible number of non-stiff loops for performing loop analysis. If no event has been generated, this computation is not performed as LoopMap will remain the same.
4. Loop analysis calculates the new values of loop currents from the loop currents in the previous iteration using the new values of source voltages as inputs. However, with the LoopMap being updated after the change in the state of conduction of the circuit, the loop currents that are available from the previous iteration may no longer compatible with the updated LoopMap. The currents for each loop in this updated LoopMap need to be recalculated to ensure that the loop currents for the new LoopMap reflect the new loops. The loop currents are calculated from the branch currents and the LoopMap. If no event has been generated, this computation is not performed as the loop currents will remain the same as LoopMap has not changed.
5. Using these loop currents in the updated LoopMap, the matrix equation for the loop analysis is generated. This matrix equation is a combination of simple algebraic equations and ordinary differential equations. If no event has been generated, the matrix equation for performing loop analysis will not change and there is no need to recalculate them.
6. The computations from this step onwards will be performed whether an event has been generated or not. The voltage sources and capacitor voltages in the circuit are updated as they will be the inputs to the loop analysis. Loop analysis is now performed by solving the matrix equation that represents the loop equations of LoopMap. From the loop currents obtained from loop analysis, the branch currents are updated using LoopMap. These currents will however only be the currents in the non-stiff branches of the circuit.
7. To update the stiff branches of the circuit, nodal analysis is applied. Nodal analysis as before uses the voltage sources, capacitor voltages, and the inductor currents as inputs in order to calculate the node voltages. Following this, the currents in all branches of the circuit have been calculated.
8. The simulator now executes any control code specified by the user. The control code could change the values of parameters of elements of the circuit or may provide gating signals to nonlinear elements such as Switches.
9. The simulator now updates every element in every branch in the circuit based on the branch currents calculated and control feedback. For example, an Ammeter current would be equated to the branch current. Or a Diode may turn on if it

is forward biased or may turn off if the current through it becomes negative. For controlled elements, a Switch may turn on if it is forward biased, receives a gating signal turning it on and it conducts current in the forward direction while it may turn off if it is reverse biased, receives a gating signal turning it off or conducts a current in the reverse direction. If in this update, any element experiences a change in its parameter, it raises an event flag for the branch in which it belongs.

10. The simulator updates the time instant for the next simulation time. This update in the simulation time instant is a combination of the time step for the differential equation solver and the time events generated by the user control functions. The entire process repeats at the next simulation time instant.

It should be obvious that the above process is fairly computationally intensive. However, this circuit simulator is specifically targeted toward power electronics applications where transitions are frequent. Moreover, unlike conventional analog circuits which operate largely in the active region, power electronic circuits operate in either the cut-off or the saturation regions. Therefore, a typical characteristic of power electronic circuits is that devices change their resistance from milliohms to megaohms and vice versa. Unless the circuit being simulated in fairly simple, simulating the transient of a power electronic device is generally pointless as these require very small simulation time steps in order to capture the dynamics of devices during the transient. The focus of this simulator is toward analysis of large systems with multiple power electronic converters. With this objective in mind, the above-described simulation flow was crafted. The simulation flow has evolved over a period of several years as the simulator was continuously updated to be able to deal with more complex circuits. Most of the functions have been added as their need has arisen. However, to speed up the simulator and decrease computational burden, redundant functions and procedures have been eliminated. The process will continue as the simulator continues to evolve and subsequent description of the process flow may be slightly or even significantly different.

8.8 Conclusions

In power electronics circuits, the power devices operate either in the cut-off region or the saturation region and not in the active region. This results in multiple branches in the circuit transitioning from extremely high resistances to almost zero resistances and vice versa due to the conduction state of the devices. This poses a challenge to most circuit simulators that have to determine the conduction state of power devices. In this chapter, nodal analysis has been shown as an effective technique in determining whether a power device will conduct or stop conducting when an event occurs in the circuit.

Nodal analysis is based on the concept of short nodes and KCLBranchMap described in Chap. 6. Short nodes are nodes that are connected by branches with only wire elements and therefore essentially are at the same voltage. KCLBranchMap describes how the branches are connected between the nodes in the circuit and thus forms a quick look-up table to generate the matrix equation for performing nodal analysis. The chapter describes how nodal equations are written by using a sample circuit as an example. The first sample circuit is a passive circuit where nodal analysis is performed to determine the current through the Voltmeter.

The most evident usefulness of nodal analysis appears in solving nonlinear circuits. The chapter describes with a simple buck converter how loop analysis fails to determine the freewheeling of the inductor current. However, nodal analysis can be used to examine the effect of all the sources in the circuit—voltage sources, capacitor voltages, and inductor currents to determine which device will turn off and which device will turn on. It has been shown with the example of the single-phase voltage source inverter how the inductor current can not only cause freewheeling of a Diode but can prevent a Switch from turning on though it has received a gate pulse. Nodal analysis is therefore a powerful tool in simulating nonlinear circuits.

The core philosophy of the circuit simulator is adherence to fundamental circuit laws—Kirchhoff's Voltage Law and Kirchhoff's Current Law. A number of approximations have been made at different stages of the simulator in order to ensure stability and achieve a solution. However, a blatant violation of the two Kirchhoff's laws has always been avoided. Loop analysis ensures that Kirchhoff's Voltage Law is obeyed and that inductor currents are computed by successively integrating their di/dt. This ensures that current through an inductor will not experience discontinuities. Nodal analysis applies Kirchhoff's Current Law to ensure that an inductor current if possible always finds a path to flow thereby ensuring that a discontinuity does not occur. Capacitors on the other hand are modeled by integrating the current that flows through them to compute their voltages. Therefore, capacitor voltages could experience discontinuities if precautions are not taken in circuit design to prevent current spikes through them. However, from a power electronic point of view, this is a design consideration rather than a simulation requirement.

The next chapter will conclude this book and present the reader with the future scope of this project.

Chapter 9
Conclusions

9.1 Advantages of the Simulator

In this final concluding chapter of the book, it is time to highlight the advantages that the simulator offers. The book has described different aspects of the simulator in all its chapters. The book was written out of need to document the circuit simulator and also to share the learning experience that has been gained out of the development of this simulator. For power engineers, the number of books that deal with circuits and systems from a simulation and programming point of view is limited. It is hoped that this book serves as a good base for young power engineers who want to learn how to simulate power electronic circuits, even if they choose to use another simulation software, by the mere fact that it provides a exposure to circuits from a programming perspective.

The biggest advantage of the simulator is that it is free and open source. Most commercial simulators for power applications are fairly expensive. Students typically have access to several simulation software as universities maintain licenses and these licenses for academic purposes are usually more flexible in their terms of use. However, upon entering the workforce, an engineer in a company will have access to usually a few select simulation software as industrial licenses are prohibitively expensive. The worst affected by increasing prices of software licenses are smaller companies particularly in developing countries, and this in turn has led to another problem—software piracy. By providing a software that is completely free, the objective is to provide a long-term platform for engineers to be able to simulate circuits.

An open-source software leads to aggressive community development by talented developers. This is quite evidently the case with Python and many of the modules and frameworks built from Python. As examples of scientific software based on Python, NumPy, SciPy, and Matplotlib are the most significant achievements. By releasing Python Power Electronics under the open-source banner, the objective is to encourage community development in the energy sector. The software Web site contains a number of sample systems that have been simulated. Other open-source

S. V. Iyer, *Simulating Nonlinear Circuits with Python Power Electronics*,
https://doi.org/10.1007/978-3-319-73984-7_9

software that have active communities result in collaboration between developers and sharing of code and data. The energy sector would greatly benefit from a platform where energy professionals can collaborate and share models. A free and open-source software could be the foundation for such a collaborative platform.

The user interface of the simulator has been made as simple and automatic as possible but without a graphical user interface (GUI). The circuit schematics can be entered in spreadsheets rather than drawn using the schematic editor in other commercial software. Even though it lacks the visual appeal of a GUI, this method of entering circuit schematics is extremely effective, lightweight, and universal as spreadsheet editing software are available in every operating system. The lack of a GUI decreases the computational burden on the processor which eventually leads to better resource management. The simulator also checks for errors in the circuit such as broken branches, duplicate component names, or missing jump connectors. The simulator automatically reads the components in the circuit schematics and updates the parameter spreadsheet. The user needs only to edit the parameters and does not need to add them manually. Again, this may not be as visually appealing as the dialog boxes in a GUI but for larger circuits, this leads to decreased development time as all the parameters of a circuit schematic can be updated on a single spreadsheet.

One of the strongest motivators in designing a circuit simulator was in providing a flexible platform for a user to include control functions. Most commercial software provide a wide range of interfaces for a user to design control functions, and the major advantage of using one of these software is the ability to design control in a simple manner with blocks from a library. Though a very effective technique in verifying concepts quickly without much effort, for complex control, a user almost always needs to write code. In Python Power Electronics, a user can include control functions only through Python code. For simple applications, this might be tedious, but for more complex circuits, the simulator offers a number of features that allow a user to design flexible control. The simulator allows a user to define and update time events that guarantee that a control function will be evaluated at that time instant. These time events can be completely asynchronous to the simulation time step and they can even cause a control function to be evaluated at a rate faster than the simulator circuit solvers. This enables the user to mimic control implemented on hardware. The other option would be to decrease the simulation time step to match the hardware resolution but this would increase the computational burden and the simulation time.

To serve the purpose of being a user manual to someone interested in using the software, Chap. 5 describes an entire system with a power converter and its associated controls. The description progresses with the circuit gradually building up to its final stage so as to illustrate how a simulation can be developed in stages. Parameters of the simulation and also of the circuit components have been given with the objective of simplifying the narrative. A description has also been provided on control function design with detailed control code and the significance of variables used. Each stage of the simulation carries simulation results to describe to the user the significance of simulating in stages as it results in verifying designs. The reader is encouraged to visit the simulator Web site and examine more advanced cases and connect the description in Chap. 5 with the circuit files and description provided to be able to simulate advanced circuits.

Chapters 6, 7, and 8 describe the core simulation engine. This part of the simulator is not essential for someone who wishes only to use this simulator. However, the workings of the simulator provide a great insight into the process of circuit simulation. A target audience for these chapters in particular would be any newcomer to the field of power electronics may it be a senior undergraduate or a new graduate. Specifically, the open-source nature of the circuit simulator allows a user to display the loops, branches, and the equations being solved in order to develop a deeper understanding of current paths and conduction of devices in nonlinear circuits. Therefore, besides being a software that has significant potential applications in power systems with dispersed power converters, the software could also be a learning tool for a power engineer.

The Web site of the circuit simulator contains a number of cases. These cases in general progress in the complexity of the circuits being simulated. The objective is to eventually release cases that resemble the power systems that are the eventual target of the simulator. The readers of this book are advised to visit the project Web site to examine some of the cases released in case one of them is close to what their test system is. The advantage of the simulator is that besides being a software, it is a growing project in a research area that is a focus of great attention. As stated before, the energy sector is experiencing drastic transformations and the best approach toward a reliable future is through extensive studies and documentation.

9.2 Drawbacks of the Simulator and Scope for Future Work

With the previous section describing the advantages of the simulator, it is only fair that some effort be made to address the disadvantages of the software. To begin with, this simulator lacks a comprehensive user interface. To elaborate on this, the input to the simulator is in the form of spreadsheets and command line inputs. An effort has been made to reduce the command line inputs to a minimum. The simulator informs the user about the files that need to be updated and to give the command to continue after the update. Moreover, the simulator lists out the state of the circuit and the nature in which the simulation output is written to the output data file. However, this does require a bit of back and forth from the user which is huge leap back from the user interfaces available in commercial software where the schematic editor contains all the information about the circuit. At the time of writing the book, it has not been decided whether the format of designing the circuit using spreadsheets will be replaced by a GUI. Or for that matter, whether the updating of parameters of circuit components using spreadsheets will be replaced by another interface. However, instead of a command line, a GUI may be designed to choose the schematics and confirm the component parameters. The extent of the GUI is not fully known but unifying the interface would simplify the use of the simulator.

The next part of the simulator is with respect to the control functions. It is unlikely that a purely GUI-based alternative will be designed as realizing control using code is far closer to hardware implementation. However, the design of control descriptors could be simplified particularly with respect to defining variables and connecting inputs and outputs. At present, a user will need to manually define most of the inputs and outputs which become error prone. A GUI-based option might be made available to the user to choose from available circuit components which would be the inputs and outputs so as to simplify this process and avoid errors. Another aspect of the control functions is that currently all code must be written as Python 2 programs. This is a significant restriction as very rarely is any hardware implementation done using Python code. C language is the most popular for programming microcontrollers and for other controllers, specialized software such as Very High Speed Integrated Circuit Hardware Design Language (VHDL) or Verilog are used. It would significantly reduce the development time for a power engineer if the simulator could accept control functions written in any of these languages.

The last aspect of the user interface is plotting waveforms of measured currents, voltages, and the control variables in control functions. Currently, all these quantities are written to an output data file in a particular format. The first column contains the time instants at which the storage occurs. The remaining columns contain the stored quantities with the simulator informing the user about the order in which they are stored. The user can plot these quantities against time or for that matter against each other using any plotting software such as Gnuplot. Gnuplot is a fairly powerful plotting software and also has a number of handy functions to perform minor calculations on quantities such as scaling them by a factor or plotting the difference of two columns. However, in most cases it is a command line software and for a basic user, it is an unnecessary complication. Since, the purpose of almost every simulation is to observe waveforms, the lack of a convenient interface with the simulator is a drawback. There is a plan to design a convenient interface where a user can automatically add the variables being stored to plots rather than user commands to do so. This would simplify the user interface considerably.

Now to examine some of the drawbacks of the core simulation engine. A question that has often being asked is why has the simulator being developed from scratch when there are frameworks which could have been used as a foundation. An example of such a framework is Simulation Program with Integrated Circuits Emphasis (SPICE) which is an open-source analog circuit simulator developed at the University of California, Berkeley, in the 1960 and 1970s. Many commercial simulators have been developed using SPICE as a base ever since SPICE achieved a great deal of popularity. In contrast, Python Power Electronics has been developed almost through trial and error if the reader was to go through the blog to trace the software's initial development. Therefore, the development of this simulator has largely been reinventing the wheel. The reason for this was the origins of the simulator was from the process of simulating circuits using code. These simulations needed every circuit to be modeled as equations and those to be solved. This approach is probably not the best as it does not utilize the concepts from circuit theory that was the basis of the development of SPICE. The approach in Python Power Electronics was from the viewpoint of a power electronics engineer and an understanding of nonlinear circuits. As can be seen from Chaps. 6, 7, and 8, it was an understanding of power electronic circuits and application of fundamental network laws that was used to develop the simulator. However, circuit theory has a whole host of advanced concepts that could have resulted in simplified analysis, faster computations, and greater stability of the simulation process. Not utilizing this resource may be a drawback of the simulator.

Another aspect of the simulator is the mathematical part related to the solving of equations may they be differential equation or mere algebraic equations. The method used in the solution of equations can have a major impact on the circuit simulator—accuracy, stability, and computational burden. Python Power Electronics has used a widely found ordinary differential equation solver (Runge–Kutta Fourth-Order solver) which though effective may not be the best way to solve differential equations in the manner implemented. The advances in mathematics can be utilized to improve the simulator in a number of different ways. The simulator as has been presented in this book and is available does not utilize advanced mathematical techniques. The focus has been to model basic circuit laws and ensure that they have been appropriately used to simulate circuits. However, improvements to the solver will be both desirable and necessary as larger and more complex circuits are simulated.

Finally, as a software, the circuit simulator may be far from optimal in its design. The focus of the software has been the implementation of network laws to solve nonlinear circuits. The simulator has developed over a period of years through constant testing and modifications. The effectiveness of the simulator has been assessed in its ability to solve nonlinear circuits. However, from the point of view of software development, there has been a whole array of metrics that have not been adhered to. For example, given a particular task that a function performs in the simulator, from a software assessment point of view, it would need to be determined whether the function uses the minimum number of variables, requires the minimum number of computations, and utilizes the minimum amount of memory. Moreover, the data structures used in the simulator have not been designed to achieve the minimum amount of repetition and the maximum utilization. Though a few data structures such as LoopMap and KCLBranchMap form the backbone of the simulator, there are a number of other structures that contain information about the loops, branches, and nodes. This leads to repetition of data and poor utilization of memory.

Therefore, from a software development point of view, Python Power Electronics may undergo significant redesign in the years to come. Particularly, the software will be modified to conform to a layered structure where the user interface forms the outermost layer and the simulation engine forms the core. The exchange of data between these two needs to be designed in an efficient and minimal manner to make the most of memory and computational resources. In this age of multiprocessor computers and parallel processing, the most efficient use of computational resources might need multithreading and multitasking where more than one process is executed in parallel. This will need efficient scheduling of processes and exchange of data between them to ensure that parallel processes are independent and supply data to sequential processes to ensure that data being used is not obsolete. A vast number of changes need to be incorporated to convert Python Power Electronics from an engineering project to a professional software.

9.3 Future of the Project

It should be emphasized that Python Power Electronics is not just a circuit simulator but a tool toward knowledge generation in the energy industry. The objective is to be able to generate a continuous flow of published material in the field of power electronics and distributed systems. To combine publications with an open-source software and case studies would result in a useful information bank for energy professionals who can combine the results of case studies with the theory and analysis presented in publications. With the simulator being free and open source, licensing will not be a cause of concern and it is hoped that this would result in a broader use of the software including in developing countries.

With respect to the simulator as a software, some of the improvements as described in the previous section will be made. A basic GUI will be designed for the user to select circuit schematics and parameters, to design interfaces for control functions, and to view plots of waveforms to variables stored in the output data file. The simulator has been currently developed using Python 2. However, Python 3 is now fairly well established. Compatibility with newer versions of Python is imperative to ensure that the latest operating systems that have Python 3 installed by default can still be used without installing legacy software. The software will be redesigned to improve memory management and code will be more efficient. Multithreading to increase the speed of execution will be implemented. With respect to the core simulation engine, research will be done on advanced network analysis techniques that can potentially simplify the circuit analysis and make them more stable. With respect to the numerical solution of equations, advanced solvers will be implemented to improve stability.

Since the focus of the simulator is to be able to simulate distributed systems, a major component of the implementation will focus on control algorithms. As an example, in a microgrid with multiple power converters, the closed-loop control performance of one power converter could have a major impact on the operation of another power converter. As has been explained in the book, the simulator has an extremely versatile control interface for implementing control algorithms with a level of accuracy equivalent to a hardware implementation. This feature will be used to examine the stability and performance of distributed systems which is a cause of concern with the demands on power quality continuously increasing. For example, an industrial power system designed for a set of machines may become inadequate if the machines are retrofitted to improve their performance.

A major focus of the project is toward renewable energy in power systems. To address the power quality issues that are being raised by utility operators, the simulator will be used to simulate detailed models of the power system with renewable energy-based generators. To examine every transient, the simulator will model every control loop associated with these renewable energy generators and their associated power converters. Moreover, detailed models of power system and associated utility equipment will be developed to examine the response of the power system to renewable energy generators. The objective of these studies is to be able to examine interconnection issues related to renewables and design viable solutions to be able to ensure reliable operation of the power system.

The study of electrical machines will be essential to complete the study of power systems. These electrical machines will be power system equipment such as transformers and voltage regulators or will be industrial machines such as motors and generators. As stated before, the performance of machines has been greatly enhanced by improvements to their design and power converters being used as interfaces. However, the improvement in performance comes at the cost of greater sensitivity and a tighter regulation on power supplied to these machines. The design of voltage stabilization equipment to prevent disruption of the operation of machines will need extremely detailed simulations.

This book has been written for electrical engineers and would need a background higher than a senior undergraduate to fully understand the concepts of circuit simulation. However, every effort has been made to present concepts visually and with examples rather than with detailed mathematical analysis. This book therefore would make a rather weak academic work. The objective of this book is to introduce engineers to the details of circuit simulation independently and not as an accompaniment to a university course. Though the book would not make light reading, it would not need constant lectures and tutorials to be able to fully understand it. In the similar manner, all the publications intended with the circuit simulator will be generalized as far as possible to be able to reach a larger of audience of energy professionals. As stated before, the complete goal of this project is to generate a continuous flow of information related to the energy industry. It is hoped that this project would also be of use to practicing engineers who would not have the leisure to completely follow the advances made in academia.

Bibliography

1. Lutz, M. 2009. *Learning python*, 4th ed. Limited: Shroff Publishers and Distributors Pvt.
2. Kaura, V., and V. Blasko. 1997. Operation of a phase locked loop system under distorted utility conditions. *IEEE Transactions on Industry Applications* 33 (1): 58–63.
3. Ghosh, A., and A. Joshi. 2000. A new approach to load balancing and power factor correction in power distribution system. *IEEE Transactions on Power Delivery* 15 (1): 417–422.
4. Kazmierkowski, M.P., and L. Malesani. 1998. Current control techniques for three-phase voltage-source pwm converters: a survey. *IEEE Transactions on Industrial Electronics* 45 (5): 691–703.
5. Mohan, N., T.M. Undeland, and W.P. Robbins. 2002. *Power electronics: converters, applications, and design*, 3rd ed. New York: Wiley.
6. Valkenburg, M.E.V. 1974. *Network analysis*, 3rd ed. Prentice Hall College Div.

S. V. Iyer, *Simulating Nonlinear Circuits with Python Power Electronics*,
https://doi.org/10.1007/978-3-319-73984-7

MIX
Papier aus verantwortungsvollen Quellen
Paper from responsible sources
FSC® C105338

If you have any concerns about our products,
you can contact us on
ProductSafety@springernature.com

In case Publisher is established outside the EU,
the EU authorized representative is:
Springer Nature Customer Service Center GmbH
Europaplatz 3, 69115 Heidelberg, Germany

Printed by Libri Plureos GmbH
in Hamburg, Germany